P9-DYY-851

THE SECRET LIVES OF GLACIERS

≈ ≈

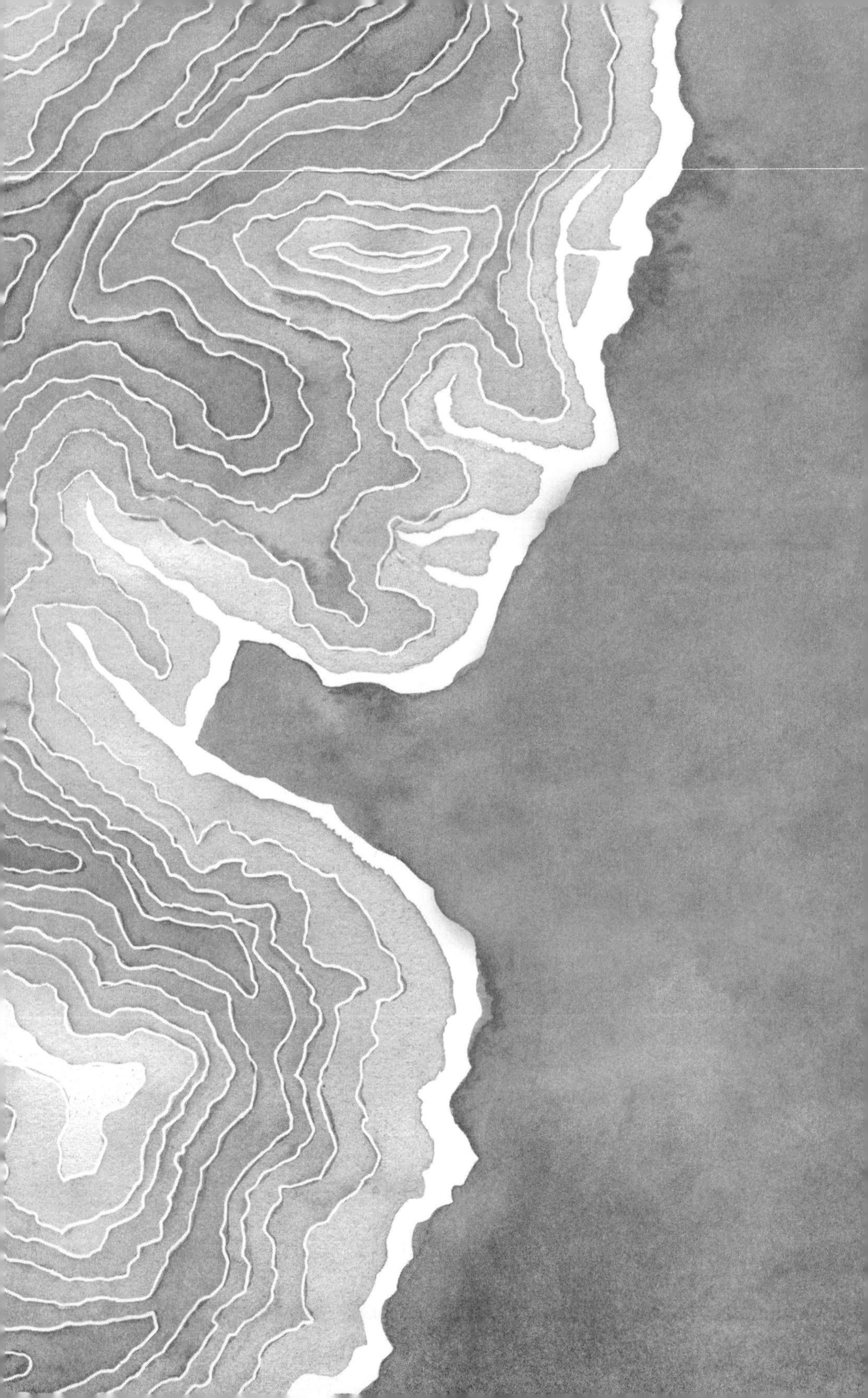

THE SECRET LIVES OF GLACIERS

≈ ≈

M JACKSON

GREEN WRITERS PRESS | *Brattleboro, Vermont*

Printed in the United States

10 9 8 7 6 5 4 3 2 1

Green Writers Press is a Vermont-based publisher whose mission is to spread a message of hope and renewal through the words and images we publish. Throughout we will adhere to our commitment to preserving and protecting the natural resources of the earth. To that end, a percentage of our proceeds will be donated to environmental activist groups. Green Writers Press gratefully acknowledges support from individual donors, friends, and readers to help support the environment and our publishing initiative.

Giving Voice to Writers & Artists Who Will Make the World a Better Place

Green Writers Press | Brattleboro, Vermont
www.greenwriterspress.com

ISBN: 978-0-9962676-7-0

COVER DESIGN & MAP: ANI PENDERGAST
BOOK DESIGN BY HANNAH WOOD, DEDE CUMMINGS & GRANIA POWER
COVER ILLUSTRATION: LAURA MARSHALL

THE PAPER USED IN THIS PUBLICATION IS PRODUCED BY MILLS COMMITTED TO RESPONSIBLE AND SUSTAINABLE FORESTRY PRACTICES

We always did feel the same
We just saw it from a different point of view
Tangled up in blue

— Bob Dylan, 1975

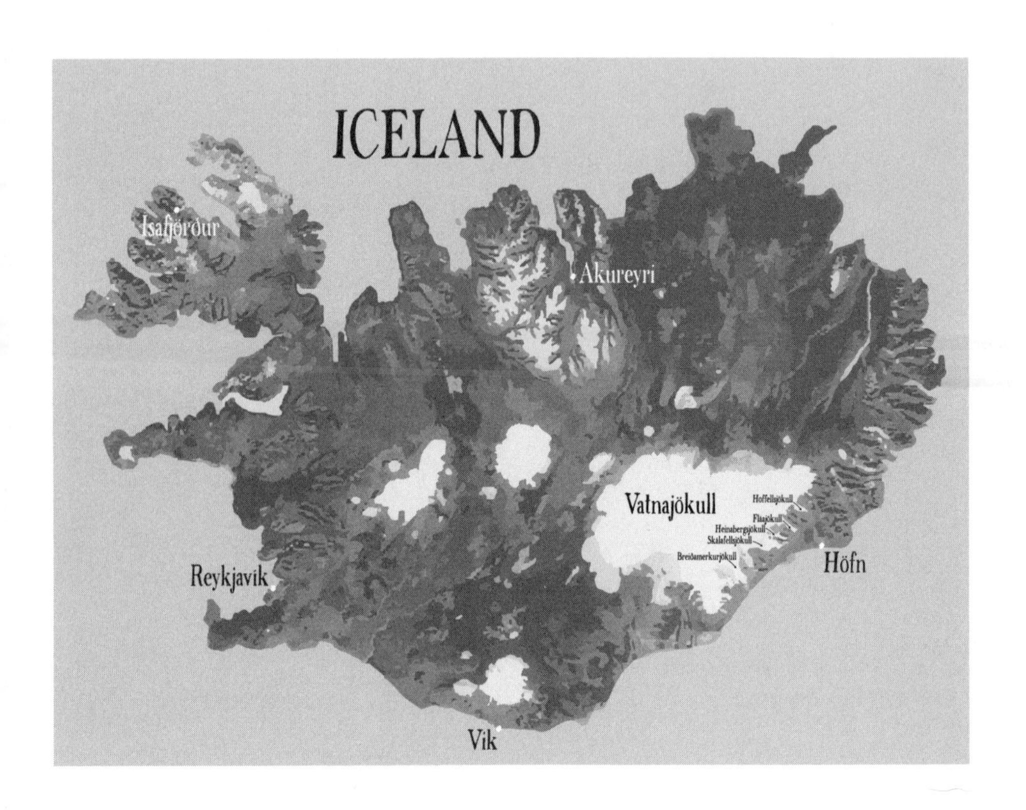
ICELAND
Ísafjörður
Akureyri
Vatnajökull
Hoffellsjökull
Fláajökull
Heinabergsjökull
Skálafellsjökull
Breiðamerkurjökull
Höfn
Reykjavík
Vík

To Jon Marshall, for the days together, the days apart, and all the days to come.

CONTENTS

≈ ≈

ACKNOWLEDGEMENTS

BOOKS ARE THE PRODUCTS OF TIME, of minutes and hours and days continuously given. I'm grateful to those who have generously given me their time, those who have poured more coffee and explained over and over, those who have shared maps and pictures and vehicles and homes and walks and meals, those who have edited and argued and encouraged, those who have supported, and above all else, those who have believed in me.

Particularly, I'd like to acknowledge and thank my team of amazing and unstoppable women who have shepherded me and this work along: publisher Dede Cummings, editor Rose Alexandre Leach, designer Ani Pendergast, artist Laura Harbage, and the entire team at Green Writers' Press.

I'm grateful for the substantial time my friends, Christine Carolan, Leslie McLees, Anna Moore, Doug Foster, Nick Perdue, Cristina Faiver-Serna, put into reading sdrafts and chapters and ideas.

Jan and Lary Marshall offered airplane tickets, home cooked meals, and support while Jeralyn Jackson, Charlie Western, Sarah

Jackson, and Kevin Malgesini handed out hot coffee, encouragement, wisdom and advice.

The U.S. Fulbright Program and the Icelandic Fulbright Commission, the National Science Foundation, the University of Oregon's Geography Department, and the National Geographic Society were all central in supporting my research.

Ford Cochran brought extensive expertise and kindness to my research. I'm grateful to the additional members of my dissertation committee for their time and insight: Mark Carey, Alec Murphy, Andrew Marcus, and Shaul Cohen.

I could not have completed this book without the immense support, time, and generosity of the hundreds and hundreds of Icelanders who shared their country with me. There are too many to list, but you know who you are. Thank you.

Thank you to the fantastic team at the University of Iceland Research Center and Nýheimar—all of whom greeted me with open doors, especially Þorvarður Árnason, Gudný Svavasdótttir, Davíð Stefánsson, Hugrún Harpa Reynisdóttir, Margrét Gauja Magnúsdóttir, Kristín Hermannsdóttir, Snævarr Guðmundsson, Soffía Auður Birgisdóttir, and Árdís Halldórsdóttir.

I gained life-long friendships throughout this research, including a sister in Sigrún Sveinbjörnsdóttir. Stephan Mantler, Inga Stumpf, Haukur Ingi Einarsson, Helga Árnadóttir, Fanney Sveinsdóttir, Asdís Jónsdóttir, and Sindri Ragnarsson helped me access the ice, provided support in the community, and poured more coffee.

This book would not be remotely possible without Jon Marshall and the unceasing support and love he offers me each day.

PREFACE

≈ ≈

ICELANDIC LANGUAGE

WHILE THIS BOOK IS WRITTEN IN ENGLISH, many words, names, and places are in Icelandic. As such, it is helpful to understand a little bit here about the Icelandic language.

ICELANDIC LANGUAGE & PRONUNCIATION

The Icelandic alphabet has 32 letters. Four letters, C, Q, W, and Z, are not part of the modern Iceland alphabet. Pronunciation is relatively straightforward. Each letter indicates a specific sound, and stress typically falls on the first syllable of each word.

Aa- as 'a' in 'man'
Áá- as 'ou' in 'house'
Bb- as 'bp' in 'spit'
Dd- as 'd' in 'daughter'
Đð- like 'th' in 'breathe'
Ee- as 'e' in 'bed'
Éé- as 'ye' in 'yet'
Ff- as 'f' in 'father'
Gg- like 'g' in 'good'*
Hh- as 'h' in 'hello'
Ii- like 'i' in 'hit'
Íí- as 'ee' in 'meet'
Jj- like 'y' in 'yes'
Kk- as 'k' in 'keep'
Ll- as 'l' in 'live'
Mm- as 'm' in 'man'
Nn- as 'n' in 'not'
Oo- as 'o' in 'not'
Óó- like 'o' in 'sole'
Pp- as 'p' in 'pot'
Rr- trilled identical to Spanish rolled 'r'
Ss- as 's' in 'soup'
Tt- as 't' in 'top'
Uu- like German 'ü' in 'über'
Úú- like 'oo' in 'moon'
Vv- like 'v' in 'very' with a light 'w'
Xx- as 'x' in 'six'
Yy- as 'i' in 'hit'
Ýý- as 'ee' in 'meet'
Þþ- like 'th' in 'thin'
Ææ- like 'i in 'hi'
Öö- like 'u' in 'fur'

*like 'g' in 'good' at the beginning of a word; like 'k' in 'wick' between a vowel and -l, -n; like 'ch' in Scottish 'loch' after vowels and before t, s; like 'y' in 'young between vowel and -i, -j; dropped between a, á, ó, u.

PRONUNCIATION NOTES

Đð (Eth) and Þþ (Thorn) are variants of the 'th' sound. Hv is pronounced 'kv.' Double 'll' is pronounced 'lt' with a click. There are additional variants. For an excellent tutorial, see the University of Iceland's free online course on the Icelandic language.

ICELANDIC PLACE NAMES

I have kept place names in Icelandic as Icelanders use them when speaking in English. Some place-specific suffixes in Icelandic denote types of environmental features such as -jökull: glacier (Hoffellsjökull/Hoffells-glacier). Others include: -fjörður: fjord; -flói: bay; -fjall/-fell: mountain; -heiði: moor/mountain; -á/-fljót/-kvísl: river; -vatn/-lón: lake/reservoir; -dalur: valley; -hraun: lava field; -sandur: outwash plain; -öræfi: wilderness.

THE SECRET LIVES

OF GLACIERS

≈ ≈

CHAPTER ONE

≈ ≈

A COUPLE OF YEARS AGO I was living on the south coast of Iceland, and one day, a man knocked on the door of my home.

He asked if I wanted to see something. No adjectives. He just asked if I wanted to see *something.*

I almost didn't hear his knock. My house was on the extreme southeastern coast of the island—literally twenty feet from the sea—and strong winds were bashing the concrete walls and making the tin roof shriek with each gust.

The man's second knock, an insistent loud pounding out of sync with the wind, drew me to the door and his question. I considered. It was cold, the wintery light was growing dim, I was a foreigner in the area, and if I went missing, no one would go looking for me for days. But then again, it was Iceland, one of the safest places in the world. And my curiosity was piqued.

I agreed, went back inside, and grabbed my nine-hundred-fill down jacket, gloves, and a hat. He had his vehicle parked right beside my door. I ran outside quickly and stepped high up into his sizeable Icelandic super jeep—the type that requires a little ladder to climb into the cab—and we drove slowly through the orderly, windblown streets of the village of Höfn.

Höfn [pronounced Hhh-Uphn], my home for several months by that point, is a small, low-lying village of about roughly seventeen hundred people built on a jagged spur of land jutting south off the island's coast like a hitchhiker's thumb. Höfn is the primary village within the Municipality of Hornafjörður, a 127-mile-long region encompassing a vast swath of Iceland's southeastern coast.

Matching glacial lagoons fan out east and west on either side of Höfn, resembling murky butterfly wings from the air. Directly south, the storm-laden North Atlantic Ocean edges the town, and to the north, glaciers pour down out of the encircling coastal mountains. Höfn—and all of Hornafjörður—is Iceland's glacier central.

A lone road led north out of Höfn and connected to country's highway: Hringvegurinn, the Ring Road, the single highway encircling the entire island. We drove for an hour west on the Hringvegurinn, then turned off the road and parked at a random moss-covered pull-off.

We both hopped out, pulled on packs and extra layers, and headed away from the road across loose rocks and thick vegetation. My host didn't say much, and wind wrapped us into quietly murmuring cocoons. Just a few clouds dotted the sky, and the light angled low, matte-gray, matching the surrounding rocky gray landscape and rising mountains ahead. Winter sunlight in Iceland tends to be low and weak but highly valued.

We gradually gained elevation over the pitted terrain as we moved away from the coast, rough loose debris bulldozed into place by decades of glaciers seesawing along the low aprons of the mountains that were once the island's coastal sea cliffs. At the top of one ridge, abruptly, the glacier Breiðamerkurjökull rose up right in front of us.

The man beside me sighed audibly in appreciation.

Breiðamerkurjökull's face—the terminus—was miles across, white but not pure white, gray and black and blue collectively impersonating white. The body of the glacier itself was lashed with thick, uneven dark moraines, ridges running tip to toe like icy tiger stripes. In the low light, the parent ice cap feeding all thirty miles of Breiðamerkurjökull, Vatanjökull, dissolved in the distance into the sky. For a moment, I was disorientated. It felt like the ice just kept sweeping vertically up into the horizon.

That's one of the hardest things about interacting with glaciers. They are often so large that atmospheric perspective—the effect where objects appear to merge into their backgrounds over large distances—distorts our abilities to accurately assess distances, scales, change. Breiðamerkurjökull is the third-largest glacier in Iceland, with a perimeter stretching over nine miles from east to west. But it is difficult to assess the entirety of a glacier nine miles by thirty, so instead, you're left with a feeling that the glacier just *dominates*.

The man and I kept a steady pace hiking towards the ice, up rocky scree slopes, and down, and back up, covering terrain in constant flux. Eventually we reached the land-ice edge and stopped briefly to put on helmets, harnesses, and crampons—spiked metal devices that strap over boots and provide traction on ice.

Moving from land to ice is tricky, as often that is where the glacier is most fragmented, brittle, and quick to break and roll in on itself, but we transitioned with little fanfare and slowly worked our way up. We wove around deep crevasses, sharp drop-offs, rock piles of debris, and stacks of wind-blown snow that had frozen into oddly-shaped pale hills. The surface of a glacier is rarely smooth; often a glacier hide is populated with shallow cuts and dips and depressions and tubes and tunnels that plummet the entire depth of the ice.

I knew we had arrived at the destination my host had in mind when he paused at the rim of a wide, shallow bowled area on the surface of the glacier. The bowl resembled a swimming pool, except there was no water because it was a glacier, and winter, and everything was frozen and cold. We had moved upwards on the ice, but not too far inward, and the mountains rose to the east only about a half mile distant.

We climbed carefully down the steep ice slope into the bowl. I'd been to this place before—for years as a geographer and glaciologist, I'd been researching glaciers and people all along the southeastern coast of Iceland, and I'd spent a great deal of time on all of the area's local glaciers. But I had not been to this area with this man before, and I did not know what it was that he wanted to show me.

He'd said little the entire journey, and he didn't break his quiet once we reached the bottom of the depression. That didn't trouble me; I'd found over the years that people will get around to telling you what they want in their own good time. My dad always told me you couldn't push a river, and I found the analogy worked just as well for people.

Large seracs—towers of ice that tend to stick up like shark's fins from the surface of the glacier—rose up on the far edge of the bowled area, and jagged pillars teetered to the west and cast deep shadows over us. Chilled, I pulled on more layers from my pack. Glacier work is all about layers.

The man removed two foam mats from his pack, handed one to me, and gestured for me to sit down. He passed me a thermos of thick coffee and a plastic adventure cup, and then he started to speak. He told me we were going to sit right there at the bottom of the ice bowl on top of the third largest glacier in Iceland right before night fell and we were going to wait.

And that's what we did. We sipped coffee, listened to the wind blow and the ice pop and crack, and watched the light grow darker and darker. We waited and made a little small talk, and he told me a little about himself and growing up in the area, and twenty minutes went by, and then another twenty minutes, and then, right when I thought I was going to be too cold to stick it out, it started.

It was dark one minute in the cloudless Icelandic sky, and then the next minute it wasn't, and the northern lights, the aurora borealis, appeared in the sky above us. First a dull glow, and then, like a light switch flipped on, blazing yellows, purples, greens, swirls of pinks and whites, and—wait—the glacier we were sitting on, Breiðamerkurjökull, it began picking up, internalizing, swallowing, *containing* the lights in the sky. The northern lights pulsed through the ice at the rim of the bowl, through the thin seracs, transforming them into icy Jedi lightsabers smoldering in kaleidoscopic concentrations. And the bowl of the glacier itself, it was whirling, throwing light like a candle-lit chandelier, like a phosphorescent ocean wave, like a field at midnight populated with hundreds of summertime fireflies.

I was engulfed. I'd never witnessed a glacier aglow with the aurora—I'd never even seen a picture of it—and standing there I felt innate companionship, as I too was as lit up as the sky.

And so we sat there on that glacier in Iceland in the middle of winter and watched. We stayed as long as we could before the clouds rolled in and obscured the sky and the lights. In the last minutes, as the ice and sky grew dim, the man turned to me and said, "This is why glaciers are worth fighting for."

. N .
V ✦ A
· S ·

GLACIERS ARE DISAPPEARING IN ICELAND.

Glaciology models predict Icelandic glaciers will lose 25-35% of present volume over the next fifty years, largely as a result of global climatic changes.[1-3] How Icelandic glaciers appear today is likely to be unrecognizable to you and me in a few decades, and simply incomprehensible to ensuing generations looking through your old vacation photographs.

Iceland isn't alone: glaciers worldwide that have existed for centuries are disappearing in human timescales—our lifetimes. Collectively, we're reaching the nadir of planetary glaciation.

Disappearing ice holds staggering consequences—after all, glaciers grow worldwide, in the Arctic and Antarctica, along the Equator, in the Middle East and central Africa. Today, we have over 400,000 glaciers and ice caps scattered across Earth, over 5.8 million square miles of ice.[4, 5] Each glacier is exceptionally diverse, each fluctuating in multitudes of complex ways to local, regional, and global environmental dynamics.

Glaciers have always fluctuated, but never at the rates experienced today. Yes, there have been times when the planet has had less ice, and times when the planet has had more ice, but—and this is a huge but—never before in human history has ice worldwide decreased as quickly as it has over the last several decades. The planet's current ice loss is unprecedented, and substantial evidence tells us that anthropogenic climatic changes are to blame.[1-3]

This unhappy marriage between immense ice loss and climate change has led glaciers to be increasingly recognized as one of the most visible icons of global environmental changes.[6, 7] Images of glaciers, data about glaciers, and stories of melting glaciers are ubiquitous throughout media and academia and popular culture, serving as visual and tangible evidence that climatic changes *are* happening and that people *are* transforming earth's systems.[8] And, taking this a step further, by virtue of glaciers' relationship with climatic changes, what I have observed is that more and more, the line between the two blurs. Glaciers act as veritable proxies, stand-ins, for climate change.[6] As in, to measure, assess, or explain glaciers is to also measure, assess, or explain climate change. So when I say a glacier has melted X amount, I'm also saying climate change has altered something X amount.

Talking glaciers is talking climate change, because glaciers are icons of climate change.

As well-known geophysicist Henry Pollack's observes: "Nature's best thermometer, perhaps its most sensitive and unambiguous indicator of climate change, is ice . . . Ice asks no questions, presents no arguments, reads no newspapers, listens to no debates. It is not burdened by ideology and carries no political baggage as it crosses the threshold from solid to liquid. It just melts."[9]

But I have researched glaciers for decades worldwide, and I can tell you glaciers are *not* just thermometers, and ice does not "just melt."

Herein then lies the problem. Glaciers more and more are iconized with climate change, but as this transpires, everything else about a glacier seems to fall away.

Another way of thinking about this is that the more glaciers melt, the more the immense diversity and complexity of glaciers also melts away. What we see happening is that today, glaciers are increasingly reduced, simplified, and detached from environments, from people, from socio-political-cultural processes. And instead, glaciers are known more for their single association, their single stereotype, their single story of climate change.

We recognize today that single stories are problematic. As Nigerian writer Chimamanda Ngozi Adichie thoughtfully observed, "The single story creates stereotypes, and the problem with stereotypes is not that they are untrue, but that they are incomplete. They make one story become the only story."[10]

Single stories are easy. They spread like wildfire. But single stories are composed of half-truths, fragments, small bits that masquerade as the whole. Single stories transform into social fact quicker than you summon the energy to even consider fact-checking.[11-13] And while Adichie was referring specifically to single stories of culture, people, and Africa, I have found that single stories are also problematic for ice.

Single stories of glaciers bolster perceptions that all a glacier can be is its melt. That's like saying that all a person can be is their gender identity, or their immigration status, or even their death.

So I'll tell you, there are no single stories of glaciers.

Consider: everywhere glaciers are located on this planet, they are located within inhabited and historic environments. Where there are glaciers, there are people (even in Antarctica!), and the two have been interacting for the entirety of human history. It is almost unimaginable in the face of the dazzling diversity of human beings across this planet throughout time, and the immense geographic diversity of glaciers, that we know ice today largely through a single story of melt.

For instance, the glacier I was sitting on watching the northern lights, Breiðamerkurjökull, has receded over four miles since 1890.[3] Since the 1970s, Breiðamerkurjökull's recession rate has increased, and approximately two miles of ice length at the glacier's terminus has vanished, leaving behind a sizeable glacial outwash plain, Breiðamerkursandur. These are the numbers—four miles of recession since 1890, two miles of recession since 1970.

But that's just part of the story. The place where I sat watching the aurora borealis set the glacier aglow was once, upon Iceland's settlement well over a thousand years ago, vegetated meadow and birch forest. Early Norse settlers built farms in the area, raised turf buildings and sheep and goats and children until around 1600 or so, when Breiðamerkurjökull began advancing over those homes and stock and children and futures.

As Icelandic families fled before the oncoming ice, elsewhere in the world colonists were establishing Jamestown in Virginia, Galileo Galilei was doubting Earth's centrality in the solar system, and the final touches were put in place on the Taj Mahal. Once Breiðamerkurjökull started to surge, the glacier crept so far forward it nearly reached the sea, stopping only 300 meters short of the North Atlantic Ocean. During that time, glaciers all over island grew larger and larger, and Icelanders by the thousands were dying in ordinary horror from cold, plagues, poverty, volcanic eruptions, and colonialism.[3]

Breiðamerkurjökull oscillated back and forth, back and forth, and as it moved, it contoured the lives of those who lived in its shadow. As early as the 13th and 14th centuries, Icelanders were writing about their glaciers in the Sagas of the Icelanders, stories of glaciers that gave ice human emotion and destiny and Icelanders a sense of their own identity as Icelanders.[3, 14] Some Icelanders even argue now that the origins of glaciology as a science didn't start in mainland Europe as is traditionally taught;

rather glaciology began in Iceland predicated on the continual connection and exposure amongst people and ice.[3, 15]

That's all part of the history of this glacier. Fast-forward to 1890, and Breiðamerkurjökull began to recede, and Icelanders started repopulating the area, releasing sheep to graze newly exposed pastures. Off-island, the Wounded Knee Massacre unfolded in South Dakota, automobiles and planes were assembled for the first time, Wilhelm Röntgen uncovered X-Rays, and Arthur Conan Doyle brought Sherlock Holmes to life.

And now, another century is gone, the planet has fully entered the Anthropocene—where humanity's impact on Earth is so serious that society has declared a new geological epoch—and the glacier maintains a backward march towards gone, dissolving so quickly local Icelanders fear it might never stop and Breiðamerkurjökull will disappear completely down to the last snowflake.

There are diverse stories stretching across human history with just this glacier, with just this landscape, with just these people within the context of human history—just as there are with every other glacier and people worldwide.

Fast-forward again, all the way to today, and early glacier stories resonate throughout Breiðamerkurjökull's neighborhood in surprising ways. While glaciers in Iceland are noticeably receding, I found this information does not necessarily *mean* glaciers are melting. That even though, as Icelandic glaciologist Helgi Björnsson observed, "the average retreat of glaciated areas has increased from about 10 to 30 km^2 per year from the mid-twentieth century until the last two decades,"[3] the metrics of change do not match the experience of change. I found that Icelanders often interpreted glacier change first through the lens of their own cultural stories about glaciers.

For example, as this book explores, older Icelanders tell stories of immense destruction at the hands of surging glaciers, and speak of feelings of safety and relief that the ice is now receding and can no longer harm them. Other Icelanders point to centuries of recorded stories of local glaciers growing and shrinking and say that these stories are evidence that, while glaciers today may be receding, this is not particularly evidence of climate change in Iceland. These people emphasize that the glaciers will return in a decade or so. Some other Icelanders tell me stories of glaciers alive, breathing, watch ing, waiting. Younger Icelanders tell stories of how they are returning to the southeastern coast to work with the thousands of tourists drawn there each year by the island's receding glaciers. These Icelanders struggle to make sense of individual profit they're making from glacier recession, of how they can learn to navigate short-term financial benefit in the face of a much larger phenomenon of change.

UNEQUIVOCALLY, glaciers are not just melting bodies associated with climatic changes. I've consistently found that glaciers are profoundly entangled with people, with individual and community lifeways—and that glaciers influence human societies as much as human societies influence glaciers. And I've come to realize that it is vital that we begin understanding these deep connections between people and ice; it is these connections that shed light on the harrowing complexity of being alive today during this time of immense environmental and social change.

It is also my opinion that disregarding the diverse stories of change opens the door for climate change deniers to selectively argue for single places in time where glacier change isn't occurring at the same rate as in other places (that one glacier is growing!) or possibly benefits local communities over shorter timescales (glacier melt gives farmers more pasture!). Coordinated denial arrayed with cherry-picked examples could dismantle years of progress on climate change momentum within the public arena.

Ignoring these stories also risks simplification of the human experiences of climatic changes themselves. As geographer Neil Adger observed when he wrote, "society's response to every dimension of global climate change is mediated by culture,"[16] how people perceive change is filtered first through their own culture, not rational understandings of scientific facts or geography.

As a scientist, it took me a long time to realize that standing up and repeating the metrics of glacier change was not very effective. That the metrics were simply just a few notes in the larger melody of change. People are not static; they rarely adhere to a single defining story of glacier change, human change, climate change—rather, like the environment itself, stories always change.

This is a critical reminder that context is *always* crucial. While it is important to have the best physical data, statistics, and models chronicling glacier change, if such information is not grounded within the human stories of glaciers (or of rivers, forests, or backyards, etc.) then that information is powerless. If people do not see themselves in the story, then they are not part of the story.

I believe that in order for all of us to realistically move forward with needed mitigation, adaptation, and transformation strategies to engage with our rapidly changing world, and to get diverse peoples worldwide to understand climatic changes and participate, it is critical to begin paying attention to the complex stories people tell about their changing environments—their changing backyards. It is these stories that often determine what people can and cannot see, and how people fit themselves into larger processes at play.

Today, as anthropogenic climatic changes intensify throughout the troposphere—the layer of atmosphere where you and I live and where climate happens—I am starting to suspect that the questions of critical importance no longer center on *if* the planet's glaciers (or tundra, forests, deserts, swamps, gardens, etc.) transform. Rather, the important questions center on *how,* and to *what effect.* And *who will people be without ice*?

In essence, the questions are no longer of mechanics, but of processes. And thus far, largely due to the continued advancement of a single story of glaciers, we have been unable to answer in any substantial depth the question of what glacier change might actually mean in real life to a region, a community, a person, you, or me. Glacier change (and most change) is rarely experienced quantitatively.

We know very little about interactions between people and ice through time, largely because we haven't looked. While some researchers have examined glaciers through societal issues including risks and hazards,[17] indigenous knowledges,[18-20] spiritual,[21-23] economic,[24] political,[15, 25, 26] gender,[6] and scientific perspectives,[7] the overwhelming majority of research investigating glaciers over the last three centuries has been from the perspective of the physical sciences—and it has focused on the physics and chemistry of glacial ice, impacts of glaciers on landsystems, and issues of ice mass reduction, recession rates, and climate modeling. In other words, on melt.[6] The information derived from these studies is of substantial value and often produced through decades of hard work by large teams of researchers. However, the produced information often keeps within the well-worn groove of the single story of glaciers-as-barometers, glaciers-that-melt. This research tells us quite a bit about the mechanics of glaciers, but cannot tell us much about everything else. It doesn't tell us what happens to *people* as glaciers melt. It is, essentially, only one chapter within a much larger book of ice.

Occasionally people are included in this single story of glaciers, but it is usually within the context of melt and a specific, predicted outcome that melting glaciers might entail for people and communities: destruction.[7, 27, 28] For example, glacier melt has been associated with, respectively, the destruction of the entire global system, threatening the livelihoods and lifeways of 1.5 billion people in the Himalaya region, and as a viable parallel to peoples' experiences from the 2001 terrorist attacks upon the World Trade Center in New York City. To some, this is what glacier melt is. This is the conclusion to the single story of melt, the single story where climate change happens, glaciers melt, and people suffer.

We hear this story over and over because our social imaginary is saturated with two things that have today become closely merged. First, immense glacier loss. Second, immense negative impacts of modern climate change.

The line of thinking often goes like this: glaciers are melting, glaciers are melting because of climate change, climate change is universally negative, glacier melt is universally negative, people immensely suffer. Oversimplified, yes, but at its bones this is the foundation for how glacier change is widely perceived.

But just as no single story exists of a glacier, and glaciers do not just melt—glacier change itself is neither all negative nor all positive.

I do not deny the seriousness of glacier melt, nor its exacerbation by anthropogenic forces, nor the unassailable reality that people in Iceland and the world across are indeed losing their ice—and there are and will be many grave impacts as a result of this unprecedented loss. Climatic changes and global glacier recession have created and will create complex problems for many people in many places over many time periods.

At times, I return to glaciers I've researched for years to find them unrecognizable and withered, and this breaks me.

But what I am advocating for is that today's single story of glaciers has not been, and is not today, representative of the entirety of the life of a glacier. The life of a glacier that includes individuals, communities, culture, scale, negative and positive impacts, geography, place—in short, complexity. Complexity that has no single story, no presupposed value, no certain future. I have been privileged to get to know many glaciers across this changing planet, to sit down and listen to diverse interpretations, stories, and knowledges of glaciers told from hundreds and hundreds of perspectives by people who live near and far from ice. I have listened to countless stories of glaciers, and witnessed profound complexity. Especially in Iceland.

I have traveled in and out of Iceland and the Municipality of Hornafjörður for over nine years working for the National Geographic Expeditions, and in 2015-2016 I lived in Hornafjörður's main village of Höfn for a year to specifically research people and ice while finishing my doctorate degree. Hence, the diverse stories in this book about people and ice are the compilation of years of experience living and working in Iceland and formally interviewing and talking with hundreds and hundreds of Icelanders.

Hornafjörður is a thin strip of coastal land stretching along the southern belly of the island, beset by wind, low clouds, thick fog, occasional trees contorted at their bases, skittish reindeer and intermittent Arctic foxes, vibrating calls from the Common Snipe (*Hrossagaukur*), and continuously shifting intensities of light. In some places, the low arable land of Hornafjörður is less than two miles wide, bracketed by steep escarpments of volcanic strata cleaved by glaciers, and harbor-less coastal outwash plains tattered by North Atlantic storm surges. From the main village, glaciers are visible from almost every structure in town and are dominant forces in the surrounding rural countryside. Generation after generation of Icelanders have continuously lived in Hornafjörður since Settlement, with isolated Icelandic families eking out marginal livelihoods farming, harvesting fish from the sea, and negotiating the regions' oscillating glaciers. Today, thousands of tourists arrive each day to experience the area's landscape and interact with ice, and locals are finding their icy neighbors objects of global attention and sources of lucrative income.

As glaciologist Helgi Björnsson wrote, "Nowhere in Iceland has the proximity and relationship between man and glacier and its rivers been so intimate and difficult than along the coastline south of Vatnajökull."[3] As one local man told me, "My father used to sweep the glacier from his door each morning."

Because of all of this, Hornafjörður is a perfect place to push back against single stories of melt, to instead deeply explore people and ice. To bring to the fore the rich diversity of both how people interact, interpret, and narrate glaciers and the rich diversity of glaciers themselves. Just as no two people in the world are ever the same, neither ever are two glaciers. In this vein, the chapters in this book are organized around each of the five dominant glaciers surrounding Höfn.

First, there is a chapter that situates the reader in Iceland—this is not a book about the history of the island, but it is essential to understand some historic events to better contextualize people and ice today. Next, five chapters move through distinct glaciers, grounding physical data and individual responses to climatic changes within complex and contradictory stories of ice told by individuals and communities. Hornafjörður itself actually has more than five glaciers—the island's largest ice cap, Vatnajökull [Vat-na-yerkotl], drains over thirty outlet glaciers down into the municipality. But from Höfn, five glaciers clearly dominate east to west, and the book's chapters follow this pattern: Hoffellsjökull [Hof-fells-yerkotl], a glacier pivotal to the Icelandic history of glaciology and demonstative of the connections between Icelanders and ice; Fláajökull [Flawl-yerkotl], which historically released many devastating and powerful glacier floods upon the community; Heinabergsjökull [Hay-na-bergs-yerkotl], a glacier that has responded to climatic changes in unique ways and propelled diverse comunity interpretations of glacier change; Skálafellsjökull [Skowl-la-fells-yerkotl], a glacier that embodies different perceptions of livingness; and Breiðamerkurjökull [Bray-the-merk-ur-yerkotl], a glacier not visible from Höfn but, through tourism, has had profound economic impacts on the village and region.

The chapters in this book offer another way of looking at glaciers and people, of rediscovering what has been recognizable to Icelanders living with glaciers over the last thousand years. Paraphrasing historian Simon Schama, this is not another explanation of the ice we are losing; rather, it is an exploration of what we may yet find with ice.[29]

My intent here is to shift the narrative needle on how people worldwide think about glaciers, and to arouse greater consideration of the complexity and richness amongst ice and people that varies by place and time. I want to bring meaning to why a quiet Icelandic man would knock on my door and take me several hours to his favorite place on the south coast in freezing temperatures to explain to me what he was fighting for—what to him was at stake as we hurdle forward into an unknown, warming future.

Years ago, Canadian anthropologist Julie Cruikshank asked, after decades researching indigenous people and glaciers in Alaska and Western Canada, if glaciers were "good to think with."[19] This book answers Cruikshank's question, emphatically responding with a "yes!" grounded in distinct glaciers, individuals, communities, cultures, scales, geographies, and place. A "yes!" supported by the belief that in order for all of us and our environments to make it through this time of immense transformation—the Anthropocene—we need to begin thinking *with* our glaciers, our rivers, our local landscapes, and our environments.

I believe the time has come for new ways of telling stories about glaciers—and for listening to the stories glaciers tell us.

CHAPTER TWO

≈ ≈

"I CAN SEE IT, why you'd want to live here." My friend and co-worker Jes Therkelsen nodded as he spoke, gesturing generally at the village of Höfn sprawled out behind us.

It was 2011, and Jes and I were taking a break from work and sitting on one of the wooden benches dotting the paved walking path edging the village's coastal perimeter. We were in Höfn for a couple of days to dry off and do laundry, a rest break in the middle of one of the National Geographic Student Expeditions we were leading. A talented filmmaker and photographer, Jes taught the high-school-aged students from around the world photography while I shored up their environmental education with Iceland-specific glaciology, geology, and climate science.

While we sat on the bench, our students were spread out across Höfn working on the on-assignment tasks we'd given them. Some students were trying to tell the story of Höfn through a series of images, photographing the single grocery store, Nettó, or the two banks, a handful of restaurants, hotels, and clothing shops, or the tiny medical clinic and sprawling golf course. Such few buildings made many of the students observe that Höfn seemed like a remote outpost, a clutch of buildings clinging to a few prominent sea rocks sticking up out of the North Atlantic Ocean.

Other students focused on more residential areas of the village, photographing the brightly painted houses, locals' laundry lines, lawn ornaments (including the ubiquitous rusty fishing bobbers that had time and again washed up on the coast), or the bent, wind-abused trees within the only two couple-hundred strong village "forests."

Iceland was deforested over a thousand years ago upon initial settlement, when Norse settlers axed the small birch forests and other native vegetation that covered at that time about a fourth of the island. They consumed the forests for construction, heating, and creation of agricultural land, and they raised free-grazing flocks of domesticated animals, primarily sheep. In the early centuries, Icelanders didn't fence their animals, and every time vegetation would take root, animals grazed through and nibbled it down, effectively stopping any resurgence.

The students, so used to Iceland's largely treeless landscape, shrieked when we arrived in Höfn and saw the patches of fenced forest. I urged them to look through the trees and see the massive glaciers cascading down through mountains encircling half the village. The students stared.

Jes and I were actually sitting on the bench because of that ice I'd pointed out. Once we sent the students off into Höfn, we'd grabbed coffees and ambled along on the coast path. We had walked for about a half mile, dodging runners, dogs, prams, and tourists alike before selecting a bench that faced the ice full on.

From our seat, we faced mostly west, and the immediate view was dominated by Höfn's westernmost lagoon, Hornafjorður.

Water surrounds Höfn on three sides, with large distinct lagoons to the east and west and the North Atlantic Ocean to the south. The Hornafjorður lagoon is a glacially fed, shallow, sediment-choked basin chocked with brownish glacial outwash and a scattering of low-lying islands fuzzed with grass. The eastern lagoon, Skarðsfjorður, is fed by small rivers and streams and, while equally shallow, is a clearer blue color. Both lagoons wrap around Höfn in a quasi-heart shape that about broke my own heart the first time I saw it from the air. Where the lagoons meet at the heart's sharp tip is also where saltwater from the ocean enters and infiltrates both lagoons during high tide.

Just beyond the lagoons, Jes and I could make out narrow strips of land dotted with clusters of buildings; just barely discernible in the shadows of the coastal mountains rising high, jagged escarpments of volcanic debris carved into blade-thin ridges. Dominating the entire scene from our bench, however, were the glaciers—the ice cap Vatnajökull and tongue after tongue after tongue of outlet glaciers lapping down through and over and around the mountains. Höfn essentially is curtained by ice.

I'd brought along a map on our walk, and Jes unfolded it and anchored it on the bench with our coffee cups and thighs. His faded, green-canvassed leg kept the North Atlantic Ocean controlled, and I pinned down the Snæfellsnes peninsula. Using long blades of grass as pointers, Jes and I touched different glaciers, trying to match the names on the map with what we were seeing from our bench.

"The biggest is the icefield Vatnajökull," he said, "but we miss most of it from here."

On the map, Vatnajökull—what glaciologists term the Vatnajökull Regional Glacier Group—spread over much of the southeastern portion of the island like a pale many-armed octopus, instantly believable as Europe's largest glacier. It looked enormous. But from our bench, the icefield appeared to spread out in a thin line behind the much closer mountain peaks. Low clouds blurred the streak where ice met horizon, and the icefield domed slightly away from us.

The tentacle-like outlet glaciers made much larger impressions. They were closer and visibly enormous.

Jes and I worked our way west along the map, slowly identifying each glacier we could see from the bench. Facing Vatnajökull, the easternmost glacier visible to us was the Hoffellsjökull/Svínafellsjökull glacier complex. Marked as just Hoffellsjökull on the map, the ice flow used to be two separate glaciers. However, since 1890, Svínafellsjökull retreated much more swiftly and today exists indistinguishable from the primary flow of Hoffellsjökull.

The map showed that Hoffellsjökull wasn't actually the easternmost—the glacier Lambatungajökull took that honor, though the mountainous terrain hid the ice and it wasn't visible from our vantage point. Just to the west of Hoffellsjökull, another hidden glacier was invisible from Höfn—Viðborðsjökull, a diminished glacier tucked quite far up a narrow valley. Continuing west, our grass blades outlined the next three visible glaciers, Fláajökull, Heinabergsjökull, and Skálafellsjökull. Heinabergsjökull sat so far down in its valley depression that it was just barely detectable.

That was all the ice we could see from our location, but, because that morning we'd driven from the western side of the island, we decided to continue moving down the map. We traced the glaciers pouring down into Hornafjörður even though we couldn't see them from our bench. After Skálafellsjökull came Birnudalsjökull, Brókarjökull, and Fellsárjökull. These three glaciers were similar to Viðborðsjökull, tucked far up into remote valleys and hard to access. When years later I hiked up to these glaciers, I'd be exhilarated to find them each distinctive and just as wild as they looked in local images.

Continuing west, the next glacier Jes and I saw was Breiðamerkurjökull, the complete opposite. While Breiðamerkurjökull wasn't visible from Höfn, it certainly overlooked the village. Several miles of its terminus front ran along the highway, and such proximity created many tourist attractions, including Jökulsárlón, glacier trekking, and ice caves—all of which provided economic benefits to local people throughout Hornafjörður and ensured that many people likely interacted with this enormous glacier more than any other ice on the island.

All the glaciers Jes and I outlined resided within Iceland's Vatnajökull National Park. Established in 2008, the national park covered the entire Vatnajökull ice cap. In the years following park conception, many surrounding areas were added to the national park, and by 2017, 14% of Iceland's total land area was within the boundaries of the park. Skaftafell, a small farm on the westernmost edge of Hornafjörður, today acts as the bustling headquarters for Vatnajökull National Park. Höfn hosts a satellite park visitor center in the recently renovated building Gamlabúð, and the village itself serves as the de facto hub for the entire 125-mile region's government, fishing, agriculture, and tourism sectors.

Jes and I had camped the previous few nights with our students in Skaftafell, and as we made our way east along the Ring Road, we threaded through Hornafjörður like a needle, stitching together the smaller districts, farmlands, rivers and bridges, and continual cascading glaciers.

By that time, I had traveled to many of the world's glaciated regions, and I had been studying glaciers in Alaska for years.

But I had never laid eyes on a place like the area surrounding Höfn—where people and ice lived so closely together. Glacier after glacier after glacier lined up like white teeth in an enormous icy grin along the highway, some glaciers so close you could see their textured crevasses near their termini. Hornafjörður was effectively sandwiched between two dynamic forces: ocean to the south, mountains to the north. In the middle, the thinnest shaving of land stretched along the coast just barely above sea level, and intrepid engineers had constructed a road to unite the entire region.

Farm settlements were labeled on our map with names like Smyrlabörg, Hnappavellir, Fagurholsmyri, and Hof, clustered between ridges of basalt, glaciers, glacial rivers, sandurs, marshlands, and other natural phenomena. Each settlement—comprised of numerous farmhouses, outbuildings, pastures, and potentially a church—was in former times an island isolated unto itself, separated from neighbors by hazardous rivers, glaciers, and ocean. Farmhouses and glaciers were in some places separated by less than a mile.

Over the centuries, the farm clusters on the southeastern coast developed their own communities, and eventually, their own farm districts that from west to east were named: Öræfi, Suðursveit, Mýrar, Nes, Lón, and Höfn. Höfn was not a farm district per se, but as it was the region's village, it had its own district distinction. As Jes and I had slowly made our way east with the students, we looked for the small stones along the highway that were engraved with the names of each farm district's boundary before they combined in 1998 into one larger region: the Hornafjörður municipality.

Feasting with my eyes on all that ice through the vehicle's windows, I felt like I was coming home. I wanted to know all about those glaciers, what it was like to live with ice here—and how Icelanders had managed to survive for hundreds and hundreds of years since Settlement. I wanted to know what was happening now that the ice was openly receding, and what impacts it might hold for this community.

"I want to live here," I told Jes that day sitting on the bench. "I want to understand what it's like to live so closely with all this moving ice."

"You'll do it," he told me.

Four years later, I arrived in Höfn with two stuffed duffle packs, two laptops, and two grants from the US Fulbright Commission and the National Science Foundation. I'd taken a bus from the island's capital, Reykjavík, all the way to Kirkjubæjarklaustur, and then transferred to a minivan for the second half of the journey. I was the only passenger, and my bags took up a full second row of seats.

The driver, Magnus, a 200-pound, well-over-six-foot-tall man who could barely squeeze into the driver's seat, had just moved back to Höfn in the last year, seeking retirement and quiet after living for several decades in the capital. We chatted back and forth as he drove, and he asked me quite skeptically why I was moving out to Höfn.

When I told him glaciers were the draw, he immediately nodded his head in understanding. "You know," he told me as we drove the Ring Road three hours outside of Höfn, "when Icelanders first came to Iceland, this whole area was green. The ice, it was hiding far back up in the mountains, and it didn't come down and bother them until much later."

I nodded, not yet aware that this was a story about glaciers that I would hear different iterations of from hundreds and hundreds of Icelanders from all walks of life.

TO UNDERSTAND GLACIERS IN ICELAND, it is first necessary to understand a little bit about Iceland's natural climate. Iceland's landscape is noticeably distinctive from other Arctic environments. Bounded entirely by water, the island centers at 66 degrees north, sharing latitude with Arctic cities including Alaska's Nome, Greenland's Nuuk, and Sweden's Umeå. But go west to Greenland or east to Scandinavia, and the climatic conditions are *much* harsher. Iceland, contrary to its name, is in all respects quite mild—and it is this mildness that helps grow glaciers.

Average temperatures range from 35 degrees Fahrenheit in the north of the island to 39 degrees Fahrenheit in the south. In the highlands, the interior of the island (40% of the island's landmass), there are much more harsh extremes. Winter temperatures average 20 degrees Fahrenheit and summer temperatures rarely rise above 60 degrees Fahrenheit. In Hornafjörður, January temperatures hover around freezing and are regularly accompanied by rain; July temperatures range from 60-75 degrees Fahrenheit with few storms.

Even though, in all respects, the island is fairly mild, it is the intense weather variability that can be exceedingly dangerous. As locals regularly report, temperatures may plummet in just hours, and snow, rain, wind, and dust storms surge with little advanced warning. Add to this intense natural variability—glaciers, volcanoes, geysers, rock formations, lava fields, light, etcetera—and there is a built-in recipe for danger.

Sensitive to how foreign this extreme weather and environmental variability might be to foreigners, in 2010 the Icelandic Association for Search and Rescue launched safetravel.is, a service that through text messages, social media, and other features provides safety alerts and other safety information to travelers. I found the service invaluable when I was out working on local glaciers and received text messages alerting me to strong winds or inclement weather. Iceland is conspicuously windy. In Höfn, locals note when it is not windy.

There were multiple times when I could not open the door to get out of my little blue Volkswagen Polo because the winds were so strong. Then in December 2015, for the first time in my life, I experienced a hurricane—Icelandic style—with winds exceeding 98 miles per hour. The government had thankfully issued strong wind warnings online and via text messaging, and like most of the residents of Höfn I stayed indoors and watched the storm out my window. Because there were few trees, it was hard to actually "see" the force of the wind, but when the light pole near my home displayed impressive flexibility, I finally got the gist of just how strong the winds were.

Typical for the sub-Arctic, around the middle of June the sun does not fully set, and in the winter for two to three weeks the island experiences fifteen to twenty hours of darkness. This causes a noticeable lull in social life in Höfn as people spend much more time either indoors or off-island. On the other hand, however, the aurora borealis regularly offsets winter darkness. I worked with a scholar at the research center within Nýheimar who would be conspicuously absent for most of the morning on the days after a particularly vivid light show. He was an avid photographer of the northern lights, and when I did not see him in the mornings I'd eagerly go straight to his social media feeds. The images he'd post of his nocturnal activities were jaw-dropping.

Compared to other landmasses, Iceland itself is geologically young—approximately 16 million years old—and rather homogenous: the bedrock is 90% igneous, comprised mostly of basalt.[30, 31] As I quickly learned, Icelanders of any age can recite by heart the island's deceptively simple origin story: the landmass was created when the mid-Atlantic Ridge edged over one of the planet's most powerful mantle plumes, and production (what a charming six-year-old Höfn boy identified as "volcano stuff") from the plume formed the basaltic mass that makes up the island. Owing to this, the entire island is volcanically active, with approximately thirty volcanic systems spreading out along a Y-branched rifting zone cutting north to south across the island. And while on average volcanic eruptions occur every five years, and I have been working in and out of Iceland for about a decade, I have never personally experienced a volcanic eruption. Friends call me the volcano-quasher.

Fascinatingly, what makes Iceland rather unusual is that many of the island's volcanic systems are covered by glacial ice. As in, many of those bubbling bowls of magma close to the planet's surface are covered—capped—with frozen ice. Fire and ice. The island's largest ice cap, Vatnajökull, sits on top of both the center of the mantle plume—a volcanic hotspot—and most of the rifting zone where the tectonic plates of the lithosphere move apart across the island. As such, Vatnajökull is an exceptionally active ice cap, with the hotspot and the thinning crust producing vast clusters of potent volcanoes and fissure swarms all far below the ice.[30, 31.] In the last eight hundred years, Vatnajökull has experienced over eighty subglacial volcanic eruptions alone.[32]

Alongside the volcanic variables, Iceland has immense glacier diversity. Approximately 10% of Iceland is covered with glacial ice, and there are five main ice caps: Vatnajökull, Langjökull, Hofsjökull, Mýrdalsjökull and Drangajökull.[1, 3] In 2008, glaciologists Oddur Sigurðsson and Richard S. Williams compiled and classified all the ice caps and glaciers on the island. In what I found to be thrilling detail, they identified 269 named glaciers within eight Regional Glacier Groups, "including 14 ice caps, 2 contiguous ice caps, 109 outlet glaciers, 8 ice-flow basins, 3 ice streams, 55 cirque glaciers, 73 mountain glaciers, and 5 valley glaciers."[33]

Iceland is not per se unusual in that it has significant glacial diversity. Most of the cryosphere (anywhere that is essentially frozen on the planet) encompasses diverse ice—but such diversity in glacier classification is not widely known. Glaciologists Michael Hambrey and Jürg Alean identify over 44 overlapping glacier types, ranging from niche glaciers and glacierets (small glaciers in gullies or depressions and glaciers over two years old that are smaller than 0.25km^2) to entire ice sheets (typically over 50,000 km^2) such as the Antarctic Ice Sheet.[34] Counting everything from glacierets to ice sheets, there are an estimated 400,000 glaciers and ice caps scattered across the planet—glaciers in the Arctic and Antarctica, in Europe and Scandinavia, in the Middle East and Central Africa, in South America and Papua New Guinea.[35]

Metrics can only take us so far, however. Some glaciologists I spoke with talked about how in the next fifty years many of these diverse types of glaciers are going to break up into smaller glaciers—essentially *increasing* the number of glaciers on this planet even as the mass volume of ice worldwide decreases.

The only two ice sheets in the world, the Antarctic Ice Sheet and the Greenland Ice Sheet, contain 99% of the world's ice. Ice caps (essentially miniature ice sheets) are the next step down. Language, and labeling, matters here: identification of ice features is based on size, morphology, thermal characteristics, behavior, and other factors. Ice caps typically are up to 50,000km^2

and constrained in some way by topography. Ice caps are under constant motion, with glacial ice forming and compressing outwards; the immense weight of the ice itself is what forces flow outward towards the edges. In addition, ice caps often create domed areas of concentrated glacial ice that flow away radially. An analog to this is the back of your hand. Look at how your knuckles are slightly domed, and your fingers flow away as connected outlet glaciers.

Iceland has only glaciers and ice caps, even though Vatnajökull is often referred to as a ice sheet. But it is technically an ice cap, and has 51 named outlet glaciers flowing away from its center in all directions.[33] Iceland's ice caps and glaciers are temperate, a thermal classification indicating that the glacier's composition occurs within a relative thermodynamic equilibrium.[1, 36] Said another way, Iceland's glaciers exist in a sweet spot of temperature that allows glacier ice and liquid water to exist side-by-side.

Iceland's glaciers crowd the southern portion of the island because of the dynamics of the global hydrological cycle. The country sits right where the warm North Atlantic Current meets the colder East Greenland Current and the northern reaches of the North Atlantic storm track. Precipitation-laden clouds—what one local woman called "clouds in need of a mop"—surfing prevailing southerly winds make first landfall at sea level on the island's southern coast—Hornafjörður—and promptly collide with the coastal mountains that used to be old sea cliffs. Via orographic lift, where air masses are forced to rise quickly and cool, the heavy rain clouds rise, transition into snow clouds, and deposit their snow loads on the steep volcanic mountains—which then feeds the growth of the island's temperate glaciers.[37, 38]

Geography is essential here. Look at any map of Iceland, and it is clear that the sizes of the ice caps and glaciers are linked to how close that ice is to the southern coast. There are a few ice caps and glaciers farther north—away from the south coast—but they are noticeably smaller. Mýrdalsjökull and Vatnajökull, both directly on the south coast, have annual precipitation between 4,000-5,000 mm.[37] These two ice caps essentially are fed more precipitation (read: snow), and subsequently produce bigger glaciers. The ice caps Langjökull and Hofsjökull, farther north and more central to the island, have 3,500 mm annual precipitation—and as such are quite smaller than their southern brethren. Drangajökull, in the West Fjords, with 2,000-2,500 mm of annual precipitation, is the size of some of Vatnajökull's smaller outlet glaciers.[37]

The origin story for any glacier worldwide essentially starts with snow and is therefore perpetually tied to precipitation cycles. The production of glacial ice and the advancement of glaciers are tied to three general processes. Basically, producing glacial ice requires snow, cold temperatures, and time. In Iceland, low Arctic temperatures ensure winter snow buildup persists into the summer months, and over decadal timeframes, lightweight snow compresses and re-crystallizes as minute granules. It is essential to understand that this is a slow process. Individual snow crystals transform into compressed glacial ice crystals quite gradually as the air pockets separating each granule slowly squeeze away under mounting pressure, leaving behind highly compressed ice crystals.

This process takes different amounts of time depending on global location and local conditions. In Iceland, with its high volume of annual precipitation, glacial ice can form in less than fifty years, though it often takes longer.[39]

While each glacier is different, typically glaciers in Iceland do not exceed a maximum thickness of 2,900 feet. Given local climatic conditions, Iceland's glaciers have among the highest mass balance sensitivities in the world; in other words, changes in precipitation patterns result in relatively rapid changes in the ice.[3, 40, 41] Increasing global temperatures can quickly trigger rapid responses in the island's ice over just a few years.

Over roughly the last three million years, the landmass of Iceland has been completely enveloped by glacial ice (termed glaciations) anywhere between 15-23 separate times.[42] Glaciations occur in response to global climatic shifts influenced by Milankovitch Cycles (variations in the planet's eccentricity, axial tilt, and precession), all of which determine the amount of incoming solar heat delivered to Earth's surface and associatively impact the growth or recession of glacial ice worldwide.[43] A base rule of thumb for ice and radiation: less radiation, cooler planet, larger glaciers. More radiation, warmer planet, smaller glaciers.

During the peak of the Last Glacial Maximum (LGM)—roughly 26,000 years ago when planetary ice sheets were at their farthest extent and there was a corresponding, substantial drop in sea levels—Iceland's land surface was encapsulated within a blanket of ice over 4,920 feet thick.[44] At the end of the LGM, as the climate warmed, almost all of Iceland's glaciers disappeared.[39, 45, 46]

But over 2,500 years ago, glaciers returned and slowly grew out again. Glaciologists speculate that Iceland's glaciers reached their most recent maximum extents during the Little Ice Age (approximately 1300-1900) and have been somewhat in recession thereafter.[39] Today, ice caps and glaciers come in a range of sizes across the island, and the legacy of glaciers is visible in the island's ice-contoured coves, bays, coastlines, fjords, valleys, plateaus, tuyas, mountains, and river channels.

THE DAY I MOVED TO HÖFN, Magnus, the stalwart mini-bus driver, peppered me with questions about the work I would be doing with glaciers. I told him that for the next nine months, I would be working out of the University of Iceland's Research Center within Nýheimar as an independent scientist. Nýheimar was a centrally located modern building that sat in the center of the village and encompassed the University of Iceland's research center alongside several other institutions. These include the Cultural Center, with its library, regional archives, and museum administration, and South Iceland Knowledge Network for adult education.

Magnus was excited for my work, and excited to tell me everything he knew about glaciers—which, three hours later, turned out to be an immense amount that touched on various topics all across Icelandic history.

When the timeline of glaciation is set down against the timeline of human habitation in Iceland, two details become clear. First, when the island was initially settled, glaciers indeed were quite smaller than how they appear today. Second, it wasn't until around 1300, as the climate globally started to cool with the Little Ice Age, that the ice crept out of the mountains and down into settlers' backyards. Unfortunately, just as the ice was growing, surging forward, Iceland was entering into what is widely regarded as one of the darkest periods of the nation's history.

For context: Iceland was settled sometime around ca. 870 by Norse migrants sailing westwards by way of the Hebrides and Faroe Islands. They brought slaves, seeds, animals, and household goods, seeking unclaimed arable land and by some accounts, attempting to escape unrest back home in Scandinavia.[47-49] Scholars approximate 15,000 to 20,000 immigrants arrived during the roughly fifty-year period known as Settlement,[50, 51] and while many settlers established themselves along the west coast near modern-day Reykjavík, a good number of early Icelanders claimed land and constructed farmsteads on the southeastern coast.

Likely, as Magnus pointed out, when Norse settlers first reached what is currently Hornafjörður, the coastal land they encountered was ice-free, birch-forested, and inviting, and glaciers were neatly tucked up in the coastal mountains. The whole region likely looked like paradise, and the settlers were quick to build sod homes and claim farmland.

Many Icelanders I spoke with narrated this period of Settlement as the country's most glorious time, referencing it habitually when speaking of past relationships with glaciers in the region.[52, 53] Stories abound of how Norse immigrants established agricultural settlements across the island from ca. 870 onward, convened in 930 a general assembly (the Alþingi), peacefully transitioned from worshiping Norse gods to Christianity in the year 1000, and preserved Icelandic history in a variety of genres including written stories such as the Sagas of the Icelanders (the Icelandic family sagas). The Sagas have helped Icelanders curate and maintain their history and culture.[53, 54] During this time throughout the country, early Icelanders identified and named many glaciers that lived in valleys or mountain tops near their homes. The Book of Settlements (Landnámabók) references about ten glacier place names—names that still endure today.

Alas, the golden age did not last.

From 1300-1850, complicated by the Little Ice Age (LIA), Iceland started undergoing shorter growing seasons, colder winters, and outbreaks of disease and famine. Glaciers started growing, overflowing out of their mountains and down over forests, pastures, and farmhouses. Severe epidemics of plague ravished the island's population from 1402 to 1404 and then again from 1494 to 1495; estimates suggest over half of Iceland's population perished.[49] Scholars categorize the seventeenth and eighteenth centuries as Iceland's Dark Ages, "the nadir of Icelandic life."[49 53-55]

Primarily as a means of ending clan conflict across the island, chieftains and other representatives of farmers in the middle of the century accepted the sovereignty of the Norwegian crown.[49, 56] The period that followed was one of abject poverty and isolation as taxation of Icelanders by the Norwegian crown increased, the Catholic Church gained considerable power, wealth, and land, and non-landowners became imbricated in a serfdom-like system.[57]

When the colonial administration of Iceland was transferred to sole Danish rule after the Kalmar Union dissolved in 1523, Danish royal decree stated that Danish merchants monopolized all trade with Iceland. As Iceland's main exports at the time, wool and fish, were of little value to Denmark, a period of extraordinary isolation and neglect began. Historian Karen Oslund suggests this purposeful neglect of Iceland by the Danish Crown was a form of internal colonialism, where uneven development or political and economic inequalities emerge within a nation or region and hold grave implications for oppressed populations. Other cases of internal colonialism include the Soviet Union's treatment of indigenous people, the United Kingdom's management of Wales, and the treatment of African American and Hispanic people in the United States. Oslund argues that Danish superiority over Iceland was not based upon race, such as in the

United States, but was based more on geography: the Danish viewed Iceland through a conflicted lens as a place both part of and completely removed from Europe.[53]

During this time, each decade in Iceland just grew darker. From 1701 to 1709, smallpox killed a quarter of the island's population. Almost five thousand people died during a famine from 1751 to1758 caused by LIA-exacerbated poor agricultural yields and high volumes of sea ice. Then, in June of 1783, the volcanic system surrounding Grímsvötn began erupting. This lasted for eight months at Lakagígar in southern Iceland, spewing toxic ash into the atmosphere, generating earthquakes, and dropping winter temperatures. One-fifth of the remaining population perished.[48, 49, 58] By 1800, the island had a population of just 47,000.[49, 59]

"You're American?" Magnus asked as we drove towards Skaftafell. I nodded.

"Many of my family migrated there. Well, not there but Canada. After Askja."

Waves and waves of Icelanders (especially from East Iceland) emigrated after the volcano Askja in the Highlands erupted in 1875. As I began to settle into the Höfn community, a refrain I heard often was how generations of Icelanders (often related to those still in the community) had emigrated elsewhere in response to disease, land loss, eruptions, earthquakes, etc. Those Icelanders that stayed were often spoken of in terms of "survival," and many present-day locals talked of how they had inherited the ability to "survive" from previous, hardy generations.

Höfn did not exist as a community during Iceland's Dark Ages—at best it was known as a local place to fish. It didn't appear on maps until after 1897, when a merchant shack was constructed for fishing. But some of the farm districts were in place by that time, and they suffered terribly during the Dark Ages, especially as the glaciers began creeping out over the region's thin strip of inhabitable land between mountains and sea.

Magnus pulled the minivan into Skaftafell at the national park headquarters. "We'll be here for fifteen minutes," he told me, then disappeared around the corner where a group of Icelandic bus drivers were chatting over coffees.

I hopped out and marveled at the change in Skaftafell since 2011, when Jes and I had camped at the small campground adjoining the park facilities. It had been quiet, with just a few people, and that was in the middle of the summer. Now, in late fall, with a distinct chill in the air, the campground was packed, the parking lot was full, and well over a hundred people crowded into the park's café. Clearly, the tourist floodgates had opened, proving what geographer and early advocate for the creation of Vatnajökull National Park Jack Ives had feared when he warned that Iceland's biggest danger was not jökulhlaups so much as "touristhlaups" [tourist-floods].[56]

Magnus returned, and we slowly got underway, driving down the park road towards the highway. He had to drive slowly, weaving amongst the groups of people standing out on the road. He didn't seem to mind them, and when I asked, Magnus told me that "they were good for the bank."

The intersection of the park road to Skaftafell and the highway has little fanfare, but I imagine in centuries before travelers were grateful when they finally made it to Skaftafell.

To either side of the highway, gray glacial outwash plains stretch away like moonscapes, and steep escarpments covered with flowing glaciers and ice falls ring the northern side. Skaftafell is located along the western edge of the farm district of Öræfi, the western-most farm district in modern-day Hornafjörður. It is bordered to the west by the river Skeiðará and the largest alluvial glacial outwash plain in the world, Skeiðarársandur, and to the east by Breiðamerkursandur, a smaller glacial outwash plain.

During Settlement, this region was verdant and forested, but in 1362 the volcano underneath the glacier Öræfajökull, which looks down on Skaftfell, erupted, triggering catastrophic jökulhlaups that flooded and decimated the *entire* region. Reflective of this, the Icelandic word Öræfi translates in English roughly as wilderness or wasteland—descriptive of the area's post-eruption condition as almost completely demolished. Entire farm settlements were destroyed overnight.

Larger vegetation like trees outside the modern fenced areas is scarce, both due to the eruption and ongoing erosion. Over the centuries since Settlement, erosion has been a continual struggle across the island, and the nation has lost immense volumes of rich soil over the years, with most of it deposited directly into the Northern Atlantic Ocean. On the upswing, erosion control on the southeastern coast is slowly changing in modern times, especially in the areas inside of and around the national park. Vatnajökull National Park's predecessor was Skaftafell National Park, created in 1967 and encompassing around 400-550 km^2 around the farm Skaftafell and some local glaciers. In 2008, the national park was expanded and renamed Vatnajökull National Park after the Vatnajökull ice cap.

As we drove through the old farm district of Öræfi, I kept my eyes open, and when I saw the roadside marker for the next district, Suðursveit, I pointed it out to Magnus. Two of the glaciers I worked with, Breiðamerkurjökull and Skálafellsjökull, terminated in the Suðursveit farm district. Continuing east, I pointed out the next marker, for Mýrar, a distinctive area within Hornafjörður. Whereas the previous two districts, Öræfi and

Suðursveit, unfolded along the windy national highway squeezed between narrow ribbons of mountains and sea, Mýrar opens in the widest and flattest lands of the entire region.

"Já," Magnus told me after I pointed out the second marker. "They had such a hard time here. It used to always flood, and the people had to fight the sea and the ice."

Mýrar translates in English to marshlands or wet lands. The majority of the region of Mýrar is a sizable outwash plain created by the glaciers Skálafellsjökull, Heinabergsjökull, and Fláajökull. However, as opposed to the sand-gray desolation of Öræfi and Suðursveit's Skeiðarársandur and Breiðamerkursandur, Mýrar is much more vegetatively progressed, with cultivated pastures spread out across the low-lying land.

As Icelandic glaciologist Helgi Björnsson explains, "In the 18th and 19th centuries, life in the Mýrar district was a constant battle with glacial rivers, with man the frequent loser in such a competition."[3] Historically, of all the farm districts in Hornafjörður, Mýrar was the most isolated and impoverished, with little wood materials, turf, or pastureland, and, continually since the 1300s, overwhelmed by Skálafellsjökull, Heinabergsjökull, and Fláajökull. Slowly, over the last hundred years, the Mýrar plain has been manipulated by the local farmers and state officials, who dug ditches, redirected glacial rivers, and constructed numerous dikes to incrementally control flood waters and jökulhlaups. But even though Mýrar is on the upswing economically, local people like Magnus still shake their heads when they talk of the district. The legacy of hardship remains in the forefront of most minds.

As we neared Höfn, the traffic noticeably increased from a few cars on the road to a few more cars on the road. And as we traveled farther east, the road progressively became narrower and narrower, with more single-car bridges over the innumerable glacial rivers coursing down from the mountains to the sea.

The road itself acted locally as a sort of marker of time. Many older people told me stories that they dated "before the road was connected," discussing times when they had to take an airplane to the capital. The actual bridging of the highway, the completion of the Ring Road in its entirety around the island, didn't happen until the mid 1970s. After bridge completion, suddenly Hornafjörður residents could drive west to the capital in just a few hours.

The connecting bridge—which is actually two separate bridges one right after the other over two glacial rivers—is at the border between Mýrar and the farm district Nes right before Höfn. It spans the glacial river Hornafjarðarfljót, which flows from the glacier Hoffellsjökull. Locally, the bridge is known as the "jumpy bridge" as vehicles traveling more than 15 mph can achieve noticeable liftoff due to the bridge's unique construction. One tired Höfn mother told me rather acerbically, "My kids sleep from Kirkjubæjarklaustur until we hit the jumpy bridge, then all hell comes out to Höfn."

Much of the road system on the coast was furthered by the sponsorship of the United States military. From the mid-1800s through the early twentieth century, Icelanders island-wide campaigned for independence from Denmark. Home rule was granted in 1874, recognition as a sovereign state occured in 1918, and in 1944, the island announced the formation of the Republic of Iceland. But in 1940, World War II British Armed Forces invaded and occupied Iceland, and in 1941, the United States assumed control. Almost immediately, the US military set about building several military bases, transportation infrastructure, and other modernizing changes across the island. Critically, on the southeastern coast, the military began building parts of what would later become the national highway, constructing bridges over shifting glacier rivers as they completed a small military base near Höfn called Stokknes (which later became a NATO radar station).[49]

The nature of Icelandic relationships with glaciers in the region shifted with the economic developments of WWII and the Cold War. The British and American militaries brought road vehicles, jeeps, and other motorized equipment before unseen in Iceland, and Icelanders rapidly folded this new equipment into their lifeways, modifying the vehicles to allow them access to the Icelandic Highlands and various glaciers and ice caps.[60]

As an Icelandic friend of mine summed it up while we drove across an outlet glacier in his hand-built super jeep, "The US brought the jeep to us, but we turned it upside down and improved it into the super jeep."

Allied forces left Iceland in 1946, and Iceland formally became a NATO member 1949 and signed a defense treaty with the United States. As such, the United States maintained a significant military presence on the island throughout the Cold War, but eventually withdrew forces in 2006. Currently, Iceland is the only member of NATO without a military force.

Magnus pulled into Höfn at five past six in the evening, just as the light was dimming, and he sighed loudly. As he drove through the outskirts, he pointed out where he lived with his wife, and he encouraged me to come by for coffee any time.

Höfn is a curious place. The trimmings of contemporary life are quite visible in the shiny red tin rooves, power lines, new cars and tractors, plastic-wrapped bales, social services, orderly streets, and fancy buildings. But it is a place where people still share stories from decades ago, or centuries before, when times were harder and people lived in turf homes threatened continually by food insecurity and scarce medical help, glacier surges and volcanic eruptions. Some residents in Höfn believe the village has weathered the last century quite well, and others perceive a troublesome decline.

Today, Höfn is the hub of both the region and Hornafjörður's social life. Approximately 1,700 Icelanders live in and around the village, and another approximately 600 Icelanders live farther out in the countryside. Many of the people who live in Hornafjörður live here because they are *from* here; they have family stretching back generations. I was often jokingly told that there were only five real families in the area, and, while initially skeptical, as I entered the community and waded into the dizzying web of how people were related to one another, I started to understand that such a quip was closer to the truth than I first realized.

These families could have left throughout the decades—and many individual family members did, moving to Reykjavík or Europe or the US—but many people stayed, enduring hardship after hardship, fishing and farming and storing up a wealth of information about what it means to live in this place. And now, after centuries, the region is on an economic upswing in large part *because* of the area's glaciers. That I'll unpack in later chapters.

Höfn and the area surrounding Hornafjörður tend to be demographically homogenous, with few foreigners living in the area. This is, however, changing very recently as more foreign nationals are moving to Hornafjörður to work in tourism and other sectors.

Iceland itself has a relatively homogeneous population of approximately 350,000 people—of which approximately two-thirds live in or around the urban western capital, Reykjavík.

Including second-generation immigrants, ca. 9% of the population is foreign-born. Before the 1990s, most foreigners were from Nordic countries and comprised just 1.9% of the population.[61] After 1990, citizens from Asia, Africa, and Europe increased, and by 2010, Polish nationals comprised 44% of all foreigners. After joining the EEA in 1994, the country experienced an increase in European economic immigrants.[61]

Höfn has a noticeable age gap, with Icelanders aged 20-35 largely absent from the area. This gap is explained primarily by out-migration to the capital area for education and career opportunities.[61-63] Locals consistently described Höfn as a fishing village within a farming community, and fishing historically was the staple industry and economic lodestone for the whole region. Over 200 residents currently work in the fishing industry, but that number is declining, likely due to complex factors including increased mechanization. Socially however, fishing is perceived, as multiple young people explained, to not require a university degree. As such, they observed, when young people leave to get educated in the capital area, they seldom return to the village to take it up.

The lack of young people is occurring in most villages across Iceland, and dates back to overall migration trends ongoing since independence. WWII and the Cold War propelled significant economic growth in Iceland and modernization of infrastructure across the island, and, critically, shifted population centers. After WWII, as Iceland speedily urbanized with the influx of foreign investment and technology, internal migration flows from rural villages to the capital region intensified.[64] Before WWII, Icelandic society was largely based on subsistence farming and small-scale fishing. Icelanders lived spread along the island's coastlines. The center of the country, covered in deserts, glaciers, and volcanoes, was largely uninhabitable; indeed over 80% of the country's landmass is uninhabited. But after the road system was connected and the country urbanized, the population increasingly concentrated around the capital. Industrialization of the fisheries, alongside the economic, political, and cultural transformations of Icelandic society in the latter half of the twentieth century also contributed to the centralization of the population.[62]

In the 1990s, migration to the capital and off island, especially for women, again increased. Researchers attribute this to socio-economic changes including fluctuating rural fishing industries, the accessibility of secondary and tertiary educational centers outside the capital—such as Nýheimar, the building that I was based out of as a researcher that also included an institution for adult education—and changing gender values.[63] Women, once educated, left rural fishing villages on the coast in pursuit of diversified job and education opportunities in the capital region.[63, 65] As such, villages were left with fewer young families. The Hornafjörður region—farming and fishing based—was no exception.

Just recently, however, demographics in Hornafjörður appear to be bucking the trend, with increasing numbers of women, families, and young people returning to the region, in part because of glaciers and the industry they catalyze. While fishing is still the second largest industry in the area, many locals now point to tourism—linked to glaciers—as the primary industry.

Magnus offered to drive me to my new home, and I accepted his offer. I had no idea where I was living. My housing arrangements had been set up by my colleague at Nýheimar. I pulled my phone out and recited to Magnus the address of my future home—Hagatún 1. Magnus took two turns, drove by the single grocery store and Nýheimar, and steered the minivan straight up to the door of my house.

My breath caught in my throat. The house had a tin roof, sleek gray siding, and was surrounded by a second-story wooden balcony that wrapped around its exterior and faced the lagoon. It was separated from the sea by only the village road and the walking path.

In less than a minute, Magnus deposited my duffle bags and drove off with a kind "bless bless!"

I straightaway abandoned my luggage and walked up the exterior stairs to the balcony. Directly across the lagoon I could see Hoffellsjökull, Fláajökull, Heinabergsjökull, and Skalafellsjökull. I laughed.

And when I looked down, adjacent to the walking path, I saw the small wooden bench that Jes and I had sat on four years previous. I'd told Jes I'd wanted to come back, to live in Höfn and understand people and ice. It was finally happening.

CHAPTER THREE

≈ ≈

"What would our nation be called without ice? Land?"
FRIÐA, 2016

"The glaciers, they are just a part of who I am."
SVEINN, 2015

I STOPPED, BREATHLESS, set my pack down, pulled another jacket out, and zipped it over the down jacket I already had on. It was a clear blue winter day, and wasn't actually that cold, but every time my friend Mummi and I got to the top of a ridge we were blasted with icy brittle wind blowing down off the ice cap. I was freezing.

Mummi had called early that morning to see if I wanted to join him out at Hoffellsjökull to "scout for some blue." Mummi was tall and red-headed, a characteristic he regularly enhanced by his predilection for wearing tight-fitting bright red outdoor gear. On the day of our outing, he was in black and red blocked pants, a fire-red Gortex jacket, and a red and white knit hat that sat high above his red curls. There was not a chance he'd ever disappear in this landscape.

We were hiking on the western edge of Hoffellsjökull's terminal moraine—the glacier's end moraine, a pile of earthen material plowed up and pushed by the glacier's snout to the farthest point of advance. Once a glacier recedes, the hill or ridge stands marooned like a lost lip, marking where the ice was once.

We had started the morning by driving north out of Höfn on the Ring Road, around Hornafjörður, the western glacier lagoon, across the jumpy bridge, and off the highway onto the road to Hoffellsjökull. At the Hoffell farm, we veered west onto a narrower dirt road and drove across the rutted single track that wove through the valley Hoffellsjökull had carved out to resemble the belly of a whale. Gentle, wide mountain slopes curved down on either side of us, giving a clear impression of how large and wide the glacier once was.

Mummi kept up a solid stream of one-sided conversation, telling me how excited he was to be out and looking for "blue." Blue in this area of Iceland means a lot of different things to different people, but to Mummi, blue translated to only one thing: glacier ice caves.

Over just the last five years, hundreds of thousands of tourists from all over the world have traveled to the southeastern coast of Iceland to stand in the magnificent light of a blue ice cave. A local man's photographs of the inside of glaciers went viral on the Internet several years previous, and people started showing up midwinter in Iceland demanding to be taken there. As one local woman described, "The tourists, all they ask is where can they take this blue, this blue picture they've been seeing on the Internet? It is all they want, that blue."

In what many Icelanders characterize as "overnight," an ice cave industry was born in the region.

"Ice cave" is a bit of a misleading term. It is not an underground chamber, like a lava tube, that is full of ice. Iceland has those in abundance, but they are not what most people refer to when they talk about Iceland's ice caves. Instead, the lay term describes essentially any hollow feature within a glacier. Features such as a moulin, subglacial tunnel, or stream channel that people may use to access the interior of a glacier. Brimming with water in the summer, come winter the features "dry up" and stand as solid blue entryways to otherworldly places. Ice caves occur in all sizes and shapes, but in the winter of 2015-2016, there were just two available (by guide only) on the south coast—both located within glacier Breiðamerkurjökull.

No accessible ice cave had been found yet at Hoffellsjökull, and Mummi was eager to locate one.

"If we find one, we'll pull more people all the way here, instead of sleeping and leaving or turning back at Jökulsárlón." Mummi was referring to how, while most of Höfn's 2,000 tourist beds in various accommodations were fully booked in the summer, tourists tended to use the village as just a bedroom stop. They didn't spend much time in the village, and they didn't often explore the area in depth. Rather, they headed out each day for more popular tourist attractions, such as the iceberg-choked glacial lagoon Jökulsárlón in front of the glacier Breiðamerkurjökull.

Mummi and I parked the car at a small rock-outlined clearing at the base of Hoffellsjökull's terminal moraine, and then raced each other up to the top of the moraine in under a minute. The view from the top of the moraine stuns—first-time visitors rarely have any idea what they might see as they drive closer and closer to Hoffellsjökull. The view is entirely blocked by the mini-mountain.

Once atop Hoffellsjökull's terminal moraine, the glacier's large proglacial lake is visible in its entirety. Proglacial lakes are bodies of water that build up in front of a glacier's terminus, typically surrounded and interspersed with unconsolidated materials generated by the moving glacier.[34] In Icelandic-English, proglacial lakes are referred to as lagoons.

Hoffellsjökull's lagoon was packed with thickset icebergs in various states of dissolution connected to each other by thin layers of frozen meltwater. The glacier itself stretched up and away into the northern horizon. From the moraine rim, the entire length of Hoffellsjökull was visible.

"I like this," Mummi breathed, standing in a braced position against the wind looking up glacier. "This is a good color with good weather!"

Hoffellsjökull stretched strong and bright before us. The glacier's surface ice had melted and re-frozen as water ice, creating a shiny, clear magnifying coat draped over the top. Not unlike a glacier blanketed in plastic wrap. The blue of the interior ice gleamed through the white, tantalizing within the crevasses and surface lacerations. I'd seen many glaciers across the world, but that day there was something magnificent about Hoffellsjökull's colors—translucent and deferred and raw and brisk and exactly what came to my mind's eye when I thought about Iceland in the abstract.

It reminded me of one of Höfn's most famous residents, the abstract painter Svavar Guðnason. He was born in 1909 in the village and painted the natural surroundings of southeastern Iceland from an early age. Svavar was particularly captivated by the area's glaciers and their variant shades of blue and white.[66] Discussing his 1953 painting *Thaw*, Svavar observed, "There [near Höfn] you see cold white colors, with shades of blue. The blue color makes it cold. If any color is Icelandic, it is the blue and white. Sky and Ice!" The blue and white of glaciers was, to Svavar, a particularly Icelandic color. "Those bright white glaciers" Svavar continued, "reflect the sunlight like I've never seen before. The water and the glacier throw the bright sunrays all around, the nature becomes more bright and lyrical than in any other place in the world."[66]

Hoffellsjökull that day with Mummi was entirely Icelandic—entirely blue, entirely white.

"Come on!" Mummi clapped his hands, sprang up on his toes, and my reverie broke. We were there on a mission. We were there to look for blue. We got going, hiking parallel to Hoffellsjökull's terminus.

Hoffellsjökull originates from the top of a vast volcano under a southeastern section of the ice cap Vatnajökull called Breiðabunga. The ice drains south off Breiðabunga, narrowing as it descends, and funnels towards Nýjunúpar—a towering chunk of rock clearly visible *in the middle* of Hoffellsjökull. The glacier ruptures as it reaches Nýjunúpar and flows jaggedly to either side of the mountain, effectively casting the mountain adrift within the ice. Nýjunúpar is technically a nunatak—an Inuit word for bedrock that pushes out above the surface of a glacier.

Looking at Nýjunúpar, it always seems like Hoffellsjökull is giving the mountain a close shave; each morning more and more rocky bristles are worn away, transported atop the ice on a slow conveyor of rubbly moraine. After the glacier reassembles on the lee side of Nýjunúpar in a confluence of ice, crevasses, and freshly created moraine—scraped materials pulled from the mountain edges as the glacier grinds by—it then flows for two and a half miles towards the terminus through the barely-over-a-mile-wide squeeze chute between two mountainous headlands called Gæsaheiði and Múli. Last, the ice drops over a sheer cliff 1,082 feet high and terminates snout-first into a trough of its own making.[3, 67]

Standing at the glacier's terminus, tracing the journey the ice makes from beginning to end, I sometimes felt exhausted on behalf of the glacier. For the entire journey, the ice squeezes and compresses and then expands only to compress again as it moves entirely confined through solid rock passageways.

When I visited Hoffellsjökull for the first time in 2012—years before my outing with Mummi—I accompanied a local Icelandic man named Ögri who grew up north of Höfn and worked as a scientist in the capital region. He was in Hornafjörður taking mineral samples, and he'd agreed to let me accompany him out to the ice. Ögri told me about how Hoffellsjökull was once two separate glaciers: Hoffellsjökull and Svínafellsjökull. Essentially, right where Hoffellsjökull today squeezes through the two mountains Gæsaheiði and Múli, the glacier had previously separated like a cloven hoof with two separate flows of ice curving around the mountain Svínafellsfjall.

Today, Svínafellsjökull no longer exists as an individual glacier. Since Hoffellsjökull/Svínafellsjökull's maximum extent in 1890, both glaciers started to recede. But due to local dynamics, Svínafellsjökull receded more swiftly than Hoffellsjökull. While Hoffellsjökull was confined by steep and narrow mountain walls, Svínafellsjökull had a less confining topography and had spread out over a sizable area of bedrock.

In contrast, Hoffellsjökull could not spread *out*, but could spread *down*. So the glacier formed itself into a compact cylinder of ice—for all intents, a sharp icy arrow—and carved out a deep trench into the soft bedrock. Consequently, when both glaciers' rates of retreat increased substantially in middle of the twentieth century, the thin pancake of Svínafellsjökull dissolved away rapidly and Hoffellsjökull remained cooly insulated in its bedrock coffin.

One of things I tell people over and over about glaciers is that geography is one of the most critical variables to consider when thinking about and through glacier change. Glacier retreat or advance is rarely uniform, and geography shapes how a glacier might respond to various stimuli. Because of Svínafellsjökull's geography, the glacier retreated much quicker than Hoffellsjökull, which was protected by a deep, cooling trench, denser body shape, and high mountain walls.

Mummi told me as we walked that locals don't refer to Svínafellsjökull much anymore—they call the entire ice system Hoffellsjökull. Mummi thought this was because there remains little obvious physical evidence that Svínafellsjökull once was beyond the shape of itself—the glacier imprint—pressed into the mountain. But even that icy fingerprint is eroding. Where Svínafellsjökull once was is today largely a wide river delta that Hoffellsjökull's meltwater flows through.

When Hoffellsjökull disappears, the glacier will leave behind a deep valley and lake sculpted in its exact shape. Between 1890 and 2010, Hoffellsjökull receded 2.48 miles[40, 41] and the trench the glacier carved out is currently 186 feet below sea level—estimates suggest it will be potentially much deeper once the whole glacier melts away.[3] The process of glacier trench-making is called overdeepening, and typically occurs when a glacier reaches its maximum extent and begins to dig down into the earth. Due to such acute bedrock erosion, carved trenches are often considerably deeper than other areas in the surrounding landscape.[34] Most overdeepenings fill with water as the ice recedes like fjords—deep valleys carved by glaciers filled with ocean water.

Mummi and I hiked off Hoffellsjökull's terminal moraine and headed west towards where Svínafellsjökull once was. Mummi walked with confidence. He told me he used to come out here regularly when his father was a glacier monitor—one of the local people who volunteered to report on specific glaciers year after year after year. Mummi's father used glacier checks as a reward for his young son—glacier blue *as a reward*.

We followed the jagged shoreline of the proglacial lake, climbing up steep ridges of loose materials, then back down, then back up. An artist in Höfn had captured this moraine in a series of blue abstract oil paintings that were just as dizzying to look at as these moraines were to walk in person. Mummi and I had little protection from vegetation, and the winds blasted us full force. Sand and small hunks of dust whirled in the air, pinching in between my eyelids, scouring my soft parts.

"This is now too far," he pointed out when we had hiked about an hour away from the car park. "Even if we find something, they can't get out here." Mummi was referring to how far it was reasonable to expect tourists to be able to hike to get to an ice cave. The two operating ice caves that winter had shorter hikes, as guides could drive their specialized vehicles quite close to the caves' entrances.

"No blue, no blue," Mummi assessed, "but it is nice to be here. . . . I used to be told stories by my grandfather of Hoffellsjökull's expeditions, which I wanted always as a child to go on. I imagined them, you know, I thought I was playing around the house that [it] was Hoffellsjökull, and I was Sigurður Þórarinsson or Sveinn Pálsson. I think this is why I am [studying to be] a scientist now in school, because I always wanted to be like our people before making glacier discoveries."

MUMMI INTRODUCED ME to his aunt Marín two months after we hiked out to Hoffellsjökull. Marín and I had tried to meet for a chat at Café Hornið, the one place in Höfn when I lived there that served a realistic-enough latte to remind me of home, but the café was mysteriously closed. So we walked two minutes to Nýheimar and took up positions on the two couches in the warm library. Near the geology section of bookshelves, a full-body taxidermied reindeer mounted to the floor watched us intently.

Another friend, Fríða, perched on the back of Marín's couch and periodically interjected her own opinion on glaciers into the conversation.

"One thing you might be interested in. See the flag?" Marín asked, and pointed to a man outside the library window who was raising the Icelandic flag on the flagpole.

I nodded. "Why was it at half-mast?" I asked.

"Because somebody died in our community. By lowering the flag, we are showing sympathy. And when the funeral in the churches is over, we [send the] flag up again, all the way up. The funeral is now done."

The three of us stared for a moment at the Icelandic flag as it reached the top of its pole. The flag, adopted in 1918 and put into law in 1944 when Iceland became a republic, represents the island's landscape. It has a dark blue background with a white cross inset by a red cross. The symbolism of the flag was explained to me repeatedly: blue represents sky and ocean, white represents snow and ice, and red represents volcanic activity.

"I connect with the flag," Marín told me. "The water is blue, the fire is red, and the white color is ice. Ice. When I was little, we learned the names of the biggest glaciers. [The ice caps] Langjökull, Vatnajökull, Öræfajökull, you had to learn these things in school because this was [what] Iceland is and this the story the flag tells us."

Friða leaned over the couch and Marín and interjected. The glaciers, she jokingly noted, were melting, and the flag might have to be revised.

"What should our nation be called without ice? Land?"

Friða's joke belies a serious question I was asked over and over. What will happen to Iceland as the ice, the blue, the glaciers—important parts of Icelandic identity to some—melts away?

When former president of Iceland (and the world's first democratically elected female president) Vigdís Finnbogadóttir wrote the foreword to Icelandic glaciologist Helgi Björnsson's comprehensive, four-decades-in-the-making treatise[3] on Iceland's glaciers, she opened the book with the simple statement: "Glaciers are an essential part of the Icelandic identity."[68]

Linger here: Vigdís wrote "the Icelandic identity." She was speaking about a larger identity that transcended individual Icelanders to encompass *all* Icelanders. And to her, glaciers were essential parts of that identity. Continuing, Vigdís wrote that glaciers "play such an important role in creating Icelanders' self-image wherever they are found," and that all "true Icelanders are fascinated by glaciers from childhood."[68] Certainly such an assessment would apply to Mummi.

Glaciers, in Vigdís's articulation, help (all) Icelanders locate who and where they are in life. While such a statement might not apply to all individual Icelanders—I certainly met several Icelanders who shrugged off glaciers—what Vigdís was speaking to was a greater sense of an Icelandic identity profoundly interconnected to ice, a certain sense of *Icelandicness*. Part of general

Icelandic identity included ice, and even if glaciers were not personally taken on by every person, most Icelanders recognized glaciers as somewhat integral to their overall cultural identity.

Anthropologist Kirsten Hastrup, who has researched in Iceland for decades, has argued that Icelandicness (her term)—a quintessential quality that makes up Icelanders and informs a collective identity—is indeed intensely contoured by ice-dominated landscapes: "Icelandic landscape is deeply marked by history and meaning for Icelanders . . . [and] is a vital part of the local memory and hence of the sense of Icelandicness."[52] Vigdís elaborated at the end of her foreword that Icelanders "owe a huge debt of gratitude to the scientists who have explored our glaciers, both past and present. They have brought to us vital knowledge about these white giants which will provide essential testimony in the great forthcoming trial concerning the potentially catastrophic case of global warming."[68]

The components of identity are hard to pin down. To some, identity might be encapsulated by colors on a flag or standing in front of a glacier or painting the brightest blue of the ice. To others, it might be fomented by history. Like Vigdís and Mummi, I found that many Icelanders evoked the history of scientific exploration and glaciological knowledge production when speaking about ice, revealing a discernible bond between people and ice across the region.

Said differently, part of the meaning some Icelanders extracted from ice included the nation's over-300-year-old remarkably well-documented history of scientific glaciological knowledge production about Icelandic glaciers by Icelandic people.[3, 33, 58, 69] Those Icelandic "scientists who have explored our glaciers, both past and present" that Vigdís refers to are the same scientists that Mummi spoke of—past Icelanders who achieved immense advancements in glaciological knowledge production and today have somewhat rock-star status in Icelandic culture, status that is evoked when some Icelanders see their nation's flag. For example, celebrated Icelandic geologist Sigurður Þórarinsson wrote in 1960:

> It has often been emphasized, and not without reason, that the Icelanders have both in the distant past and in later times achieved great things in the sphere of literature. It is much less known what they have achieved in the field of natural science. . . The fact is that although there is little mention of Icelandic and other Norse people in connection with history of glaciology Iceland has its history of glacier research which in earlier times was full comparable with the history of glacier research found in countries where glaciology is supposed to have begun . . . But no Nordic people and hardly any Europeans lived in so close contact with glaciers and wrere [sic] so affected by them and the glacier rivers as those who lived in Austur-Skaftafellssýsla [Hornafjörður]. It is no wonder that this district above all others the cradle of Nordic glaciology.[70]

What Þórarinsson wrote is a refrain I saw echoed across Icelandic society and from Icelandic people specifically discussing glacier science in Iceland.[3, 58, 69-71] The refrain boils down to two parts: first, that early Icelanders achieved significant—but often unacknowledged—advancements in the pursuit of glaciological knowledge, and that, second, these advancements are part of a cultural legacy to which all Icelanders today can look upon as part of who they are.

The history of Icelandic glaciological knowledge production, especially in parallel to how such knowledge developed in Europe, has been discussed in much greater detail elsewhere than I can do here.[3, 56, 58] It is important, however, to understand that the history of glacier science in Iceland is deeply ingrained with modern perceptions of ice. Just as Mummi could tell me he idolized past glacier scientists, so too did I hear this from many other people. I found it was not uncommon when speaking particularly with male scientists, glacier guides, and tourism operators that they would reference past Icelandic men and their scientific works as a way to explain what glaciers meant to them.

Of particular note are four repeatedly referenced historical figures that play prominent roles in modern Icelanders' social imaginaries of glaciers and help strengthen the bonds of collective glacier identity across the nation: Þórður Þorkelsson Vídalín, Eggert Ólafsson, Bjarni Pálsson, and Sveinn Pálsson. Given the historical context in which these early scientists lived, their accomplishments are remarkable.

During the late 1600s, Icelander Þórður Þorkelsson Vídalín began documenting glaciers on the southeastern coast of Iceland over the course of his daily life as a circuit physician. He compiled his work in *Bookcula de montibus Islandiae chrystallinis*, advanced a "frost expansion theory," reported that local glaciers advanced more often than receded, and explained how people living near various glaciers believed glaciers were created from snow versus the prevailing Danish theory of saltpetre. Modern scholars have argued that Vídalín's work was "the most knowledgeable and important thesis on glaciers anywhere in the world at the time."[3, 33, 72]

Vídalín's work was followed by Eggert Ólafsson and Bjarni Pálsson, who from 1750 to 1757 wrote a report called *Reise* that claimed, eighty years preceding Louis Agassiz's idea of glaciers growing and shrinking with Ice Ages, that in Iceland glaciers were growing larger (advancing) since the time of Settlement.[3] When Icelanders spoke to me about the history of Settlement, and how the ice was then smaller than it is today, they occasionally referenced Ólafsson and Pálsson, who to some scholars are the first scientists who sounded an alarm about how glaciers could advance with potentially negative impacts on surrounding communities.

Perhaps most famous of all four early glacier scientists was Sveinn Pálsson. In 1791, Pálsson, a trained physician working on the south coast, wrote "*Draft of a Physical, Geographical, and Historical Description of Icelandic Ice Mountains on the Basis of a Journey to the Most Prominent of Them in 1792–1794*." It is difficult to overstate the importance of Pálsson's work. As Iceland's foremost glaciologist Helgi Björnsson wrote: "He [Pálsson] used the research methods of Enlightenment natural scientists, researching phenomena in person, collecting facts, analysing [sic] and classifying data and then connecting the various pieces of information to present an overview."[3]

Pálsson also had a minimum "50-year intellectual lead on his European scientific colleagues in terms of substantiated field observations and descriptions of glaciers. From a history of science perspective, Pálsson would likely have been called the "father of glaciology."[73]

Unfortunately—and likely adding to his modern fame—Pálsson's research disappeared from history for almost a century. In 1795, Pálsson sent his manuscript to the Natural History Society of Denmark. Danish officials in turn sent the manuscript to the Topographical Society of Norway, where it then disappeared. This "loss" of one of Iceland's most important documents has shaped many modern interpretations of Danish colonial practices as seeking to suppress Icelandic intellectual advancement.[70, 73] Pálsson's glacier work "was entirely overlooked elsewhere in Europe and played no part in the mainstream of glaciological thought. Had it done so, it must surely have ranked as the most important and illuminating work on the subject written during the eighteenth century."[69]

Pálsson is celebrated in Iceland today even if he is less known beyond the island's borders. The hardships experienced by those who went to the ice for knowledge—and particularly to Hoffellsjökull, one of the more accessible glaciers at that time—continues to fascinate and inform modern Icelandic culture and identity. Perceptions of colonial neglect of science achievements (purposeful or not) are widely held, and the publications of Sveinn Pálsson, Eggert Ólafsson, and Bjarni Pálssons' research *in Icelandic* (as opposed to Danish or English) for the first time in the twentieth century has contributed to forging strong connections between Icelanders and glaciers in modern times.

After about the 1930s, European scientific interest in Icelandic glaciers grew, and numerous foreign expeditions—the ones Mummi specifically referenced—were carried out over the ensuing decades.[56, 74-76] Most European glacier-oriented scientists at that time only had access to smaller glaciers on the continent; with a (relatively) short trip to Iceland, they had access to enormous glacier complexes and intact glacial land systems. A notable expedition was in 1936 on Hoffellsjökull, made up of Swedish glaciologist Hans Wilhelmsson Ahlmann, Icelandic meteorologist Jón Eyþórsson, field assistants including Icelanders Sigurður Þórarinsson (Mummi's hero) and Jón frá Laug and Swedes Carl Mannerfelt and Mac Lilliehöök.

Hoffellsjökull was by far one of the easier glaciers in Hornafjörður to access. During the early part of the twentieth century, Hoffellsjökull did not have a proglacial lake at its terminus—rather, the glacier pushed right up the rim of its terminal moraine where Mummi and I had stood.

From approximately 1920 to about 1950, locals used to travel the roughly fifteen miles from Höfn north through the farm district of Nes to Hoffellsjökull for glacial ice. Using lorries, Icelanders would back up the vehicles against the glacier's terminus and load ice chipped off from the towering glacier edge.

The glacial ice—much longer lasting than water ice—was then transported back to the village and used to pack and cool locally caught fish. Hoffellsjökull at that time was perceived as a valuable asset in the community; while most of the fish caught in the surrounding waters were salted, fish exported to Britain (especially during WWII) were kept chilled with ice—often, glacial ice from Hoffellsjökull. But once the glacier started to recede in the 1940s, it began forming a proglacial lake at its terminus. For lorry drivers to access the ice, they would have had to first get across the growing lake. As such, it was easier to abandon the practice, a situation made more palatable as it coincided with a downturn in the fishing economy of the area.

However, the existence of the road alongside the even topography approaching the moraine made international expeditions to Hoffellsjökull logistically much easier to manage, and from about the 1930s onward, Hoffellsjökull was the site of intense international and national glaciological investigation.

This legacy of glaciological knowledge production carries through to the present day. For example, one local resident, Gunnlaugur, a twenty-eight-year-old assertive man who was born and raised near Hoffellsjökull, told me stories about the country's leading glaciologists who repeatedly came out to Hoffellsjökull throughout Gunnlaugur's young life, and the multiplying number of foreign scientists coming and going at Hoffellsjökull. Gunnlaugur observed that "the last two summers, we have groups from Italy. They are drilling into the mountains to get samples. . . . They give us samples and stuff like that. That's increasing and stuff like that, I think there is one guy [scientist] that has been coming here for like five years. He doesn't take anything, but just goes to the same spots. There are others, from US like you and from Europe, but I cannot remember what they do."

Widely acknowledged, however, is that what cemented Hoffellsjökull's role as a site of scientific inquiry was the 1936 expedition led by glaciologist Hans Wilhelmsson Ahlmann. While this expedition contributed important glaciological knowledge about glacier mass balance processes and mechanics,[39, 75, 76] what made it famous then (and today) was the utilization of local men from Höfn and the post-expedition instigation of a long-term glacier monitoring program. The creation of this program had immense and long-lasting cultural repercussions and has forged some of the strongest connections between people, ice, and identity throughout Hornafjörður.

Upon the completion of his expedition up Hoffellsjökull, Ahlmann empowered several laypeople and farmers throughout Hornafjörður to continue monitoring and reporting about local glacier behaviors. The reporting continued sporadically throughout the 1940s, but was inconsistent and losing momentum.

In 1950, Jón Eyþórsson—Ahlmann's co-leader on the 1936 Hoffellsjökull expedition—founded the Iceland Glaciological Society (IGS or, in Icelandic, Jöklarannsóknafélags Íslands). In 1951, IGS formally established a countrywide glacier monitoring program that is still in place today. The IGS is composed of women and men scientists and laypeople who participate in the production of glaciological knowledge, the results of which are regularly published in IGS's research-oriented academic journal, *Jökull*.

The impacts of IGS's annual glacier monitoring and reporting were profound. Many glaciers had a variety of place names attached to them, and this annual reportage effectively institutionalized certain glacier place names and the stories attached to those particular names. This is significant because in Iceland, castles, buildings, or monuments to the nation's past don't really exist—most structures built between Settlement and the eighteenth century were built of nature and subsequently returned to it. As Kirsten Hastrup remarks, in Iceland, "*words* (including names) remain the most significant remains of the past; words carry the message of antiquity."[52] The words, the names of a place, the names of a glacier, hold the history of a nation. When certain names of glaciers became more commonly used than others, invariably words, stories, and histories reduce or fade as others moved to the fore.

Another impact was that IGS began curating a (relatively) long-term dataset recording glacier change on the island—which over the long-term confirmed ongoing glacier recession at unprecedented rates, and formalized to some degree cultural narratives (within the Icelandic scientific community) of glaciers as early warning systems for global imbalances.[3] As Vigdís observed, it is Icelanders and their glacier knowledge that "will provide essential testimony in the great forthcoming trial concerning the potentially catastrophic case of global warming."[52]

Last, IGS's annual glacier monitoring program socially formalized particular identities to include *glacier monitor.* Not only are glaciers part of Icelandic collective identity, but individually, some people identify as monitors of glaciers through the IGS program. Multiple local people, particularly farmers and Icelanders without formalized scientific training, identified themselves in the course of conversations as *official* glacier monitors.

I spoke with a local man in Höfn, Diðrik, who had been an IGS glacier monitor for over twenty years. I did not know from the outset that he was a glacier monitor, rather, he immediately self-described as such. "I teach here and I am a monitor of the glaciers here," he told me upon introductions.

Note that Diðrik identified as a teacher and as a glacier

monitor. He did not say that he does monitoring, he said he *is* a monitor. That is part of his identity.

Diðrik continued, "When I am doing something with the glacier, like measurement, that's the science department of my head, but when I go there I feel something and that's a total different part of what I am and sometimes I feel a little connection between these two, the poet, and me, different corners of my head. But still I know I will go to the glacier every year and I measure it and I feel like a poet. The glaciers are important to me!"

What Diðrik explained is that when he measured local glaciers every year for IGS's glacier monitoring program, he felt moved by his experience with glaciers. To him, being moved is important and fosters connection. Being a glacier monitor enforces Diðrik's connection to the ice.

Many Icelanders, official or unofficial glacier monitors, expressed some variant of Diðrik's feelings. Officially, approximately forty people in Iceland measure fifty to sixty individual glacier snouts at present. While forty might not seem like a lot, in a small country of 350,000, forty is still a fair amount of people. As the former head of the program, Oddur Sigurðsson noted, "there are some farmers who do this. And there are carpenters who do this, painters and whatever, and we have the physician in Vík [a village on the southern coast], we have a plastic surgeon, and a supreme court judge among us, and Jón Eyþórsson's [the man who started IGS] granddaughter." Glacier monitors are in all strata of local society.

Oddur told me that more and more Icelanders *want* to be glacier monitors, especially in the wake of growing publicity for the program. "There have been several people who have approached me to adopt a glacier," he told me in his crowded Reykjavík office, pointing to stacks and stacks of papers about the program.

"Every time some publicity gets going, people get interested. And there was a report in *The New Yorker* by Elizabeth Kolbert, she wrote a book also, and she interviewed me also, I was telling her about adopting glaciers . . . and I am pretty sure that very few Icelandic scientists were as well publicized in the world as this was." Since Kolbert's article was published, Oddur observed that Icelanders of all ages wanted to adopt a local glacier. "They're not thinking about what we do for science, but they want to adopt the glacier to care for it. I tell them yes."

While IGS formalized glaciological monitoring in the sense of official glacier monitors who regularly report on specific glaciers, the practice of watching the ice and reporting out has been practiced in Hornafjörður for centuries by lay people. Glacier monitoring is, in many ways, a cultural practice, an active way of fostering connection amongst people and ice. Rich, detailed descriptions of local glacier activity appear in diverse historical sources including reports on resident glacier knowledge, county and parish reports (~1700-1900), land registers (~1800 to present), diaries, oral stories, photographs, and maps (~1600-present), and in present times literature, film, photographs, maps, reports, and oral traditions.[3, 41, 56, 77]

Many of these documents helped shape the foundation for this book—and people were time and again generous with lending me their family documents, images, and stories. Other Icelandic scientists have also relied a fair amount on these official and unofficial historical glacier accounts to supplement and support ongoing glaciological work today. For example, Icelandic scientists Hrafnhildur Hannesdóttír, Helgi Björnsson, Finnur Pálsson, Guðfinna Aðalgeirsdóttir, and Snævarr Guðmundsson utilized a mix of historical documents and photographs to reconstruct the glacial extent and ice volume of Vatnajökull's southern outlet glaciers including Hoffellsjökull during the Little Ice Age.[40, 41]

In modern times, glacier watching and reporting by laypeople has been especially associated with photography and time-lapse video techniques. A handful of people I befriended in Hornafjörður managed some variant of ongoing glacier photography or artwork, with many of their projects lasting five years or more.

One artist found historical images of local glaciers in the Reykjavík Museum of Photography's *archives* and had gradually been recreating the images to show glacier change over the last century. His before and after image series won international acclaim. Another man visited Hoffellsjökull every month and took a photograph of the glacier from the same location. The compilation of his images depicted profound glacier recession, and has been shown at a variety of regional and national scientific conferences about global environmental change. Another woman showed me a series of photographs she had taken of Hoffellsjökull over the last eight years. She was recreating the glacier's recession in accompanying watercolor paintings.

Many local people also engaged in glacier monitoring as part of living in the area, which included the ongoing construction and proliferation of local glacier knowledge. Gunnlaugur, the 28-year-old local man who lived near Hoffellsjökull, told me about his grandparents, who also lived near the glacier. His grandparents, he told me, drove out to Hoffellsjökull's terminal moraine each day to check on the glacier. "My grandma and grandpa, aged 85 and 87, they go there every single morning. Just go there, stay there for maybe ten minutes, and then leave. Just looking at stuff, sit in the car or go outside, if the weather is good. Don't hike anywhere, just go around the places. Then just go."

To Gunnlaugur, his grandparents' glacier vigilance was a source of identity and pride. He told me about them straight away in our second interview. Implicit in Gunnlaugur's view—and reflected in the views of many other Icelanders I spoke with—was that scientists were not the primary source of his glacier knowledge. Rather, Gunnlaugur learned about glaciers first from other local people who knew intimately about ice—from family members, friends, and neighbors, from people who monitored and watched the ice. And then that information was supplemented by science.

For an entire week in November after meeting with Gunnlaugur, I spent each morning out at Hoffellsjökull. I got up early in the dim sub-Arctic light, made a thermos of coffee, and drove through the still-sleeping village of Höfn, past the school and gas station and then down the long highway with pastures and mountain on one side and the glacier lagoon on the other. And then I turned off the paved road and onto the gravel, and I drove for ten more minutes until, my headlights bouncing in the rutted road, I drove towards the terminal moraine.

Every morning I parked in the dirt space where there weren't any other cars, put on a second layer, and then headed up the moraine. I sat on my usual rock in the dull light, sipped my coffee, and stared at the glacier.

It was movement that caught my eye the first morning. To my left, on a rock a fair distance away, sat an Arctic fox. Its tail was tucked over its feet, and it was facing me. It was grayish, hadn't turned to white yet, and it stared at me like it had been watching for a long time. I looked for the fox each morning after, but I never saw it again.

While I went out to Hoffellsjökull officially to take advantage of the unusually clear weather that week and to photograph the glacier in flickering starlight, I was really quite curious to see who else in the community interacted with Hoffellsjökull. The first morning, I thought it was just going to be the fox and me.

But then, sure enough, that first morning and all the mornings after, as I drank coffee out at the terminal moraine, I'd see coming down the potholed dirt road the bouncing twin beams of bright headlights as Gunnlaugur's grandparents drove out, like clockwork, to check on their glacier. Sometimes they smiled and chatted with me, and sometimes they didn't—but they weren't there to see me. They were there to check on Hoffellsjökull.

IT WAS LATE EVENING AT THE END OF OCTOBER, and I was having coffee and cookies with three local artists in an upstairs alcove in the FabLab, a shared artists' space within Vöruhúsið, an arts and crafts center in the middle of the village.

Lydia leaned forward and tapped her left hand on the stylish wooden coffee table at our knees for emphasis. "You have to know, understand, we are used to change. Iceland is a very young country. At one time, Iceland was just rock. Then the glaciers! Icelanders we look out the window we see glaciers, we see change. If the landscape is changing, all things are changing, and it is the change that connects us to Iceland, to Icelanders," she explained. The other two women sitting with us, Þóra and Berglind, nodded in agreement.

Amidst deconstructed pool tables and guitars, dusty boxes, art supplies, and various piles, the four of us sat on a large brown U-shaped couch backed up against floor-to-ceiling windows draped closed with black curtains. A wooden hand-made coffee table was stacked with ceramic cups, cheese and fruit platters, and three types of cookies. My digital recorder sat marooned amongst more stacked phones on the table than the number of people in the room.

It was pouring rain outside, so we'd all arrived for the meeting a little damp. Þóra was a landscape painter, mainly working with oils and watercolors. She'd lived in Höfn since 1983 and told me she'd only really started getting into art after 2010 when she retired from teaching. Her short brown bangs framed wire glasses and expressive gold eyes. The first thing she said to me when we met at the front door of FabLab was how excited she was to spend the next hour talking about glaciers.

I'd arranged to meet several local artists in the village. Paintings, photography, sculptures, and other pieces of glacier-oriented art were ubiquitous throughout Höfn; I'd noticed many of the homes I'd visited had large, glamorous pictures of local glaciers hung upon their walls. Every time I purchased groceries at the Nettó grocery store, I walked past a ten-foot-tall aerial image of village's glaciers partially obscured by stacked appliances for sale. By the village's waterfront and harbor, bulky cement blocks held upright a local photographer's images of surrounding ice. In Nýheimar, the common area's walls were arrayed with images of local glacier changes over time.

Just as the physical landscape echoes ice—for glaciers have carved their bodies into the earth, hollowed cirques and cwms and lakes, sharpened ridges and mountainsides, gouged fjords, stranded tuyas, moraines, eskers, and arêtes, and scattered boulders like marbles across striated bedrock—I was finding so too did the social landscape. The pervasiveness of glacier artwork in the village suggested to me that glaciers meant something—conveyed something—to locals, and I wanted to know more.

Hence, I decided to talk with several local artists and arranged for coffee and cakes at FabLab.

Lydia was a little quieter—she tended to wait until after everyone else had spoken, then she would insert her opinion in short bursts. Lydia said she'd lived in Höfn most of her life; she'd moved to the village when she was six years old with her family from the Vestmannaeyjar (Westman Islands) after the 1973 volcanic eruption of Eldfell. "Here there are no eruptions or earthquakes. That is why we moved here," she explained.

Berglind introduced herself last. She was born in Höfn one street down from where we currently sat. Her parents were farmers in Nes, a farm district outside of Höfn, and Berglind struggled with reading and writing throughout school with little support. She chose after school to go into athletics, and owned a gym in the village for over twenty years. But then she decided to go back to school to study art, and upon her graduation, she returned to Höfn and opened a successful studio in the village.

"This is a good place to do art here because here the colors are so strong," Þóra began. She described how Höfn sits farther off the coast so the village gets morning, afternoon, and evening light. "I think it is the most light that lasts the longest time in the country here."

"Yes," Berglind agreed. "We all paint yellow, red color, and the sunset is so long. So much color here!"

Þóra continued, "Where I live, this is my view from my window. I see the mountains and the glaciers to the west. So that is the thing I paint the most when I paint the landscape. The brightness and the blue on Hoffellsjökull in the afternoon."

"The brightness?" I asked.

"Well, the glaciers are so white and when the sun is shining they stand out, like they are getting up and coming at me."

"They turn blue when it is raining," Berglind added.

"Or sometimes grey! Or almost black, like purple black."

Lydia nodded. "I've seen it [the glacier] yellow and pink!"

"The glacier, it takes on the colors of the sky, or the sea," said Berglind, "everything around it the glacier takes. It always takes. Even as we see the glacier is getting smaller, it is still taking—"

"Yes," interrupted Þóra, "and the mountains they are darker as the lightness from the glaciers is shrinking."

"We see that with Hoffellsjökull, Fláajökull, with Öræfajökull, the glacier is always going up, more and more blue and his white head is getting smaller and smaller," said Berglind. "Ha! I ordered more blue and the art store was out!"

The women tell about how they find themselves using more blue paint to portray glaciers and less white paint. There is less snow in the landscape, barer blue ice. Berglind summarized: "this change, the white color is smaller and smaller and smaller, and we use blue color more and more more and more."

"Our blue is Iceland's blue," Lydia added.

The other two women agreed. Berglind expanded, "You will see no other blue than the blue here from the glacier. It is the real blue of Iceland. When I was in Copenhagen, I could not find it there and I had to come back."

"You must understand," Þóra directed, "you must understand the power of blue color of the glacier to us Icelanders. It is our blue, it connects me and the glacier."

"How? What does the blue mean to you?" I ask. The rain pounded the draped windows, and Þóra refreshed everyone's coffee cup. One thing I was slowly getting acculturated to—late evening coffees.

Berglind fiddled for a moment with a phone on the table. She pulled up an image and showed it to me. "In all my work, when I use blue, I am very long, I am maybe two times [longer] to make the blue light, because it is so good to work with blue light, the glaciers' blue. At first, when I first painted, I painted only in blue."

Þóra agreed with Berglind. She added, "Here, there is much color, much light, with the glaciers. There is so much blue, in the mountains, in the glaciers, in the ocean, in the sky. Here, is all blue."

Living directly on the water, with mountains, glaciers, and bottomless skies surrounding my home, I found myself regularly stopped in my footsteps as I walked about town and encountered abundances of blue from all directions. Oceanic moisture sheathes Iceland, imbuing the land, atmosphere, and culture with a scattered cellophane blue bordering on translucent. It colonizes the greens and grays and yellows as an almost watery afterthought, a reminder that the last color to appear in every language was blue.[78] It tints the pervasive fog and milky Arctic sunlight in equal measures of disorientation.

American writer Rebecca Solnit examined the color blue through a lengthy meditation, offering, "Water is colorless, shallow water appears to be the color of whatever lies underneath it, but deep water is full of this scattered light, the purer the water the deeper the blue. The sky is blue for the same reason, but the blue at the horizon, the blue of land that seems to be dissolving into the sky, is a deeper, dreamier, melancholy blue, the blue at the farthest reaches of the places where you see for miles, the blue of distance. This light that does not touch us, does not travel the whole distance, the light that gets lost, gives us the beauty of the world, so much of which is in the color blue."[79]

Blue in Iceland is the light that got lost in the world and found haven on an island in the north Atlantic. Art historian

Julian Freeman wrote that "[w]hen it comes to Icelandic art, outsiders are often struck by our artists' choice of colours, no matter whether the subject matter is representation, a mountain or a boat, or abstract. 'Cool' colours predominate, particularly all sorts of blues... it is unusual to find such bright, undiluted blues anywhere but in Icelandic art."[78]

I found many Icelanders especially attuned to different blues in the glacier. Fríða, the woman who sat with me in the library, would tell me throughout the year about expected weather based on the blueness of the glacier we could see out the window. "If the glacier turns turquoise, you can expect rain or snow, or something is going to happen physically." Dark blue, she stressed, always meant rain.

Another man, Guðjón, told me late one afternoon how when he moved to the village with his wife, who was from the area, he learned to also tell weather based on the ice. "The glaciers, they change color all the time. That I learned, since my wife lives here, and her parents live here, then I learned that the colors are always changing . . . there is always new color in the glacier. The black glacier, the blue glacier, the white glacier, the grey color, all these colors! Better put on a jacket for the blue glacier!"

Earlier, the women artists had talked of how the glacier took everything around it, including the color blue from the sky and mountains and sea. Even as the glaciers were getting smaller, they said, the glaciers were getting more and more blue.

At the end of the meeting, I asked them how the glacier could take more things even as it was getting smaller.

Þóra paused for quite awhile, gathering her words, before telling me: "I am not, in a way, sad the glacier is melting, because it is melting the blue into the oceans and rivers and back into the air and it is still here. I see my blue still all about."

Berglind agreed: "We have green water and blue mountains in the brightest weather. When it is good weather, you see the blue color in the glacier, you see blue color in the mountain, and sometime you can see it again, and then always have it."

They were losing their glaciers, the women artists in the FabLab. But they weren't losing their blue. It was pervasive, everywhere.

. N .
V — A
· S ·

GLACIERS ARE CHAMELEONS. They take on easily the colors of the sky, the pending rain, a people's culture and history, a young man's hunt for economic stability, an artist's dreams.

One Höfn man, Sveinn, told me: "The glaciers they are just a part of who I am." He did not explain his statement; rather, he simply repeated it to me when I asked for clarification.

Since Settlement, Icelanders have witnessed, documented, and represented glaciers in a wide variety of styles, mediums, and approaches through all sorts of intertwined narratives. Said differently, glaciers resonate and reveal.

As former President of Iceland Ólafur Ragnar Grímsson noted during a 2013 speech at an exhibition for Icelandic painter Johannes Kjarval: "Culture reflects the spirit of the people. Art reveals beauties that are hidden, unknown, showing the core of our being, how nature makes us human, life worth living; who we are; where we came from."[80]

Mummi wasn't just looking for blue because it provided economic value. I suspect he was also looking for blue because it was a part of him, a part of the blue that drove explorers and scientists to keep searching, part of what so many local artists keep trying to depict, to explain.

I spoke with Berglind a couple of days after we met in FabLab. She was eager to tell me, "I feel that the glacier is part of my life. And I think that people who live here all the time, the glacier are in their mind. Because you see it all day, the color, you see it, you know it, at night, in morning, all day, with not the same color! And you can always see a new glacier in color. I think that people here in Höfn, they love it."

They love it.

Just like, perhaps, Peruvian and Bolivian pilgrims love glaciers when they extract sacred cultural medicine from them in the Sinakara mountains.[81, 82] Or perhaps how Inuit find value in using glaciers as seasonal icy highways for seal hunting in Greenland.[83] Or how many different indigenous peoples derive powerful lessons from glaciers about cultural norms and moralities in western Canada and northern India.[18, 22, 23, 84] Or how some Swiss view glaciers as repositories for dead souls in limbo in the Alps.[7, 85, 86]

Glaciers creep across landscapes and culture, showing up and disappearing time and again. When we ask ourselves what actually happens when a glacier melts—who, how, what, when, does blue really matter—perhaps we'll find those answers when we look at the twisted, frayed, and immensely complex bonds that wrap people and ice together.

For a long time, it was assumed that glacier melting impacted a people's identity—and the people in this chapter support that assumption. But, of course, it is a little more complicated (it is always a little more complicated). It is not as if a cultural identity will go away as the ice does. That is not how identity works. Glaciers are a part of who I am—but if the ice goes away, my identity does not disappear. Rather, like a landscape responding to glacier change, my identity will respond, transform, shift. As our identities always have.

Right now, there is plenty of evidence to suggest that glaciers are part of the cosmology of Icelandic identity, that part of being Icelandic is to have some sort of connection with glaciers. Flags help shore up that relationship. So does the history of glacier exploration in Iceland. So does color. Berglind thinks the people of Höfn love the glaciers.

Here's the thing with chameleons. They aren't the thing they take on. They are not flags, history, or color. They are just themselves, and entire ecosystems entangle around them.

CHAPTER FOUR

≈ ≈

"At Haukafell children from that farm could be in their house and throw stones at the glacier."
SILJA, 2015

"You can also imagine with how much it shrunk since maybe 1,000 years ago . . . when people were settling here, just how large and intimating, just imposing across the country, the glacier was SO big. So that was like a fundamental part of just living here, at the beach, because all the towns are just at the beach because the glacier was just fucking there, you couldn't go anywhere and the glacier was just there."
VÉSTEINN, 2016

"Those floods."
DAGFINNUR, 2016

IMAGINE FOR ONE MOMENT you had a farm in Iceland a hundred years ago. Your farm was on the southeast coast, in Mýrar, the farm district squeezed in between Suðursveit to the west and Nes to the east. Looking south from your farm, you could see the dark blue of the North Atlantic Ocean, and to the north, the light blue of the sinuous glaciers Fláajökull, Heinabergsjökull, and Skálafellsjökull uncoiling through coastal mountains contoured by the ice cap Vatnajökull. You tended sheep, maybe a milk cow, your family and crops, and fell each night into bed exhausted, lulled by the sounds of crowded sleep and the damp sod roof shifting. You live always at the edge, and famine, poverty, and a single mistake could shatter you. Your farm. Your family.

Imagine one late-summer morning, you step outside, and you notice the birds are silent. The birds are still there—you see them rustling in the few stunted birch trees you've fenced from the sheep near your home. But the birds do not sing down the morning as they are wont. Rather, there is silence. You look to the two sheep you've kept close to the house the last week because one has a split hoof. The animals are pressed up near the far stone outcropping that walls half the enclosure. You see their unease in the distance.

You know something is not right.

Years ago, you and your family built your home on a large bluff of rock sticking some thirty feet up off Mýrar's coastal plain. More accurately, you built your home into and atop the outcropping, using the natural rock as one of your house's four walls. You climb now above your house's earthen roof to the steep top of the rock, and you survey the land in all directions. Everything to the south, east, and west looks normal, but looking north, you notice a darkening of the flow lines radiating out from Fláajökull's terminus like dendritic engravings.

The braided glacial rivers, slipping around, under, over each other, grow darker from light blue to blood blue to black blue and you see them each expand, fill, widen.

It makes sudden sense, and you understand what you're seeing. Your hands start to sweat. A glacier flood is coming. A *jökulhlaup*.

The last time this happened, you survived, but you will never forget watching it, the almost unspeakable power emanating from the relentless glacier meltwater choked with icebergs, boulders, sediments, vegetation, debris, and your neighbor's house and sheep as it all gushed uncontrollably towards the ocean.

You estimate you have about thirty minutes, and you work immediately to gather your family, farm workers, animals, food, water, household goods, and anything else everyone can grab and get to high ground which is only thirty feet off the plain; and by the time the first waters reach you and you pray the jökulhlaup will not be more than thirty feet, thirty feet between you and losing everything, thirty feet that you know intimately because you measured after the last flood and now you realize that thirty feet is not very high at all.

Why did you stay after the last devastating flood, or the one before that? Why did you stay after each flood, and are you going to stay this time? Or will you leave, like your neighbors have long left, as your friends keep urging. The glaciers are growing, you've seen them getting bigger and bigger—you've lost quite a bit of pastureland, and you know the glaciers keep releasing more and more jökulhlaups. Do you go, or do you stay?

For hundreds of years, Icelanders living throughout Hornafjörður lived this.

During and after the Little Ice Age,[77] but especially from the seventeenth through nineteenth centuries, climate deterioration made Icelandic livelihoods exceptionally difficult along the southeastern coast: glaciers fluctuated, surging and receding, sawing over the land and grinding it like sandpaper; farms shifted and were regularly destroyed; floods appeared with little warning, drowning livestock, children, and able-bodied farmworkers. The

Little Ice Age did not bring consistent planetary cooling; rather, it brought weather and climate inconsistency and variability.

People in Iceland couldn't count on anything.

Individuals, families, young people, old people living throughout Hornafjörður—as the climatic conditions worsened, they had to make the decision to go or stay. To migrate (to where? when? to who?).

Fast forward to present day, and people worldwide face the same decisions as their climate worsens. Go or stay?

In western Africa, villagers in Niger build walls to keep the desert out, but sands creep over pastures and crops and homes as part of increasing rates of desertification.[87]

Bangladeshi agriculturalists build their own walls to keep the sea from flooding their fields and homes. But the waters keep rising.[88, 89]

Since the 1940s, villagers in the Andes dam and drain proglacial lakes to keep glaciers from destroying their communities.[17]

And today, thousands and thousands and thousands of refugees stand on the southern edge of the Mediterranean and weigh trying to cross the treacherous waters northwards in flimsy rafts.

Whatever the environmental situation, people worldwide make decisions now much like those of Icelanders in Mýrar. Go, or stay. What would you do?

There is immense banality and horror wrapped up in confronting such a decision that lingers long after the choice has been made. The massive force of glaciers—their *power*—on the southeastern coast to force decision-making of this nature left behind a legacy that permeates the social fabric of Hornafjörður today.

When older local people told me how glaciers were growing smaller, they told me this news with voices laminated with relief, over and over pointing to the budding kernel of safety they felt now that glaciers could no longer surge or flood over their homes, farms, or communities. Many told me they no longer felt powerless.

One middle-aged man succinctly observed as we stood near his office window and looked out at Fláajökull: "Já [yes], the people living here before, they were powerless to glaciers. Glaciers and volcanoes and the sea killed so many of them then. Já, and now, today, the glacier has lost all its power, has lost all its teeth."

Glacier change is never simple. No change is simple. People up and down the southeastern coast are steeped within living memories of generation after generation of Icelanders powerless before deadly ice. Glaciers have oscillated, growing larger and growing smaller, and they have at times held immense power throughout the community. Other times, they have possessed very little power. Understanding that power—glacier power—and how it influences people and ice in this region reveals the many mixed and conflicting perceptions people worldwide hold today concerning glacier change, and, more broadly, environmental change.

THE LEGACY OF JÖKULHLAUPS is conspicuous across Mýrar, where the rural landscape differs from the rest of Iceland. Mýrar is mostly an immense glacial outwash plain—a *sandur*. Outlet glaciers Fláajökull, Heinabergsjökull, and Skálafellsjökull edge the northern side of the farm district, and the North Atlantic Ocean edges the southern side. The two forces—ice and ocean—enact a powerful back and forth struggle over the terrain (what many local people characterize as a battle of the elements), with glacially-fed rivers and jökulhlaups transporting immense volumes of materials away from the ice and to the ocean, and the North Atlantic Ocean pushing much of those materials back inland. Because of this, much of Mýrar is scraped smooth along the bedrock from the base of the mountains to the sandy edge of the sea—with one noticeable exception.

Small rocky hills rise above the sandur in Mýrar, freckling the plain like braille dots, rock outcroppings eroded and craggy on all sides and shaped by centuries of glacial floods. Perched atop these hills, with commanding views of the area, are structures—farmhouses, barns, silos, pens, stables, sheds, cabins, and others.

Driving through Mýrar, one of the first things people remark on today is that all the structures are built atop hills.

Jensína, a science teacher in Höfn for eighteen years and counting, was the first to clarify the geography of inhabitance in Mýrar. "If you take Mýrar," she explained to me one day in her office, "we know for a fact that all the homes there are built on a hill because they had such floods. Always, always, the houses were built on a hill, because of course they knew. It was a way of life. The floods came, and it was so bad."

Not all rock outcroppings in Mýrar have buildings, because, as Jensína explained, problems existed concerning clean water availability. "Farms were built close to really good and clean water supply," she observed, "but there isn't a clean water supply everywhere. You know you have to build where you are completely free, where you can be completely sure there is no problems with water."

Any rock outcropping rising high enough above the sandur with clean water was covered by homes and other structures—a basic spatial recipe in Mýrar that kept people safe from Fláajökull and other local glaciers for centuries.

Smári, a farmer in his late 50s who owned a farm in the middle of Mýrar, vividly described the floods to me: "The glacier floods are not just water, they are many bits of material like a slurry that plows your home away."

Jökulhlaups—which roughly translates to "glacier run"—are not ordinary hydrological events. They originally just referred to subglacial floods sourced from Vatnajökull, but today the term refers to any sudden and large release of water from a glacier triggered by a range of mechanisms.[34] One mechanism is when a volcano erupts under or near a glacier and melts a portion of glacier ice. The ensuing glacier meltwater can be released in a torrential flood. Another mechanism is when an internal pocket of water inside the glacier—an englacial lake—escapes, or water suspended within the glacier body, alongside englacial debris, releases and floods.[90-92] Glaciers are not solid ice; they are riddled with tunnels, channels, and all sorts of other features. As the glacier morphs, the internal geography of the ice is always transforming.

More common in Mýrar, though, was the type of jökulhlaup triggered when an ice-dammed lake bursts.[93] As the glaciers grew, sanding their long necks through the brittle coastal mountains, they at times cut across existing side valleys and effectively dammed them with ice. A lake would form in the blocked valley and grow larger and larger and fill the valley. Then at some point, the lake would overflow, hydraulicly lifting some of the damming glacier ice and escaping in a sudden catastrophic release. This was so common in Mýrar that some valleys, such as Vatnsdalur, Heinabergsdalur, and others, were known sources of jökulhlaups. Floods from these valleys had their own rhythms, names, times.

Baldur, an older, reticent man, recalled to me a chilling story of a jökulhlaup in Mýrar: "One flood came late in October, and it had been cold weather in the days, weeks, days before, and water was frozen on the land," he described. "And then the flood came. And there was this old lady, describing it to me, and she said the flood came ON ICE. It came easily. It skated! And a lot of icebergs, in all kinds of sizes also came. And she said after the flood she couldn't walk to the next farm because of icebergs and the house was gone."

Baldur looked at me as recounted this story, shaking his head. A jökulhlaup triggered in Mýrar just before winter set in slid across the iced-over and slippery ground, gaining speed as it went. It destroyed and swept away Baldur's neighbor's home. "The house was gone." It vanished. Such a story aligns with Smári's definition of a jökulhlaup: it is a flood that takes your home away.

People living in Mýrar had to contend for centuries with jökulhlaups that swept homes and livelihoods away, surging glaciers that crushed homes and livelihoods, and many other sources of hardships from rainfall-based floods, volcanic eruptions, and the ongoing weather and climate variability brought on by the Little Ice Age. Such hazards contributed to Mýrar being a historically underprivileged district.[77]

Writing about the Little Ice Age expansion of parent ice cap Vatnajökull and the southern outlet glaciers around 1400, Icelandic glaciologist Helgi Björnsson vividly captured the abject devastation the community in Mýrar experienced due to glaciers.

> Eventually one glacial tongue after another began to reach beyond the mountain passes and creep over vegetated land into the valleys, and then finally down onto the lowland. Rivers that had once meandered with clear water in fixed channels, became dark and muddy and began to spill over their banks, dispersing layers of gravel and stone beneath them. Cowherds and their cattle became cut off in meadows, outlying sheepcotes became isolated, sometimes for days, and some farms had to be abandoned as they could no longer harvest enough hay to survive, their fields cut off by impassable glacial rivers.

> The glaciers grew in size, becoming thicker and longer, and sometimes dammed tributary valleys, preventing waterways from flowing onwards. . . . This damming formed marginal lakes and when they overflowed they caused jökulhlaups, which flooded the local farms and their land. In the 18th and 19th centuries, life in the Mýrar district was a constant battle with glacial rivers, with man the frequent loser in such a competition. A century ago, travellers would have found these rivers a very different phenomena than today, for then they became swollen, frequently migrating from channel to channel, either eroding away arable land or covering it with glacial till and rocks. Drifting sand and gravel from the outwash plains also damaged farmland. Jökulhlaups became an annual event, during which farmsteads with their sheepcotes, barns and stables were turned into little islands surrounded by water. Farms which had stood on the same spot for centuries were eventually abandoned or else moved elsewhere to avoid the encroaching waters. This was a district with much poverty; its farm buildings were poorly constructed, there was little driftwood, and the turf was barely usable as the peat was sandy. There was no upland grazing for sheep because glaciers now filled in the valleys above the farmsteads. . . . The terrain gradually began to sink under the increasing overburden of the glaciers, wetlands became more widespread along the lowlands, and harbours on coastal skerries became silted up and impaired. The expansion of the glaciers had blocked mountain trails, hampered journeys between different parts of the country, and prevented the upland grazing of sheep in the highlands.[3]

As glaciers grew, Icelandic families living in Mýrar lost their land and their agricultural and fishing practices. People watched as glacier rivers jumped their banks and expanded, they lost hand-cultivated pastures and roadways and farms, saw harbors vanish and marshes grow. All because of ice. Living in this area meant negotiating daily life in a landscape *dominated* by oscillating glaciers.

FROM HÖFN, Fláajökull lies to the west of Hoffellsjökull, draining down off Vatnajökull in what from the ground looks like a spectacular S-curve. Fláajökull is an active temperate Piedmont glacier roughly sixty-seven square miles in size and eight miles long. Piedmont glaciers occur when a valley glacier—confined by steep mountain walls—flows out into an unconfined area like a large plain. Typically, a Piedmont will spread out in a large crescent shape, not unlike what happens when you spill honey onto a flat table. Fláajökull descends down between steep mountains Fláfjall and Heinabergsfjöll, around a recently exposed mountain named Jökulfell (translated to "glacier mountain"), and then out onto the sandur, terminating into a proglacial lake that drains across Mýrar's inhabited plain.[94]

Like all the five glaciers featured in this book, Fláajökull's snout face ends in coalescing proglacial lakes (lagoons). As Fláajökull recedes in the future, the current lagoons will combine and backfill into a lake estimated to be over 300 feet below sea level at its deepest point.[3, 46, 95]

Fláajökull reached its Little Ice Age (LIA) maximum around 1880-1890[96] and then began receding. Recession was not linear—again, like all the region's glaciers, Fláajökull has periodically surged and receded like sandpaper grinding down a landscape. Fláajökull's back and forth action has resulted in arguably some of Iceland's most distinctive arcuate and sawtooth moraines—distinctive crescent bands rippling one after the other in the land surrounding the receding glacier snout.[97] Fláajökull is a glacier best viewed from the air.

Fláajökull advanced somewhat between 1970-1989, but since

1989 it has receded one third of a mile, exposing old moraines, drumlins, columnar basalt formations, new mountains and ridges, and altering the paths of two rivers, the Hólmsá (flowing into the North Atlantic Ocean) and the Djúpá (flowing into Hornafjörður).[3, 97] In total, between 1890 and 2010, Fláajökull has receded 1.6 miles.[40, 41]

When Fláajökull reached its Little Ice Age maximum extent, locals today are unsure whether the glacier merged on its western flank with neighboring glacier Heinabergsjökull. Many farmers I spoke with, and others in Höfn, debated this question of the two glaciers meeting with surprising intensity. Some believed it did, and some believed it did not. People pointed to different fragments of evidence.

For example, if Fláajökull and Heinabergsjökull did meet, it was likely around 1860-1870.[41, 77] And if the two glaciers did meet, the force would have effectively removed any vegetation, materials, or other features from the landscape. Imagine your thumb and pointer finger resting in a sandbox, and you start pinching and un-pinching your fingers. You'll eventually create a groove as you move push away all the top sand. If the two glaciers did meet, it would have ground away any evidence of how the landscape once was before the ice moved over it.

It is actually because of this—the lack of observable glacial geomorphological features in the area—that a local Icelandic research team concluded that the two glaciers must have merged.[40] These findings were supported by historical documents and discussed in local informational signs recently placed near the glaciers by the park service. These included, for instance, one document from 1932 where two local farmers gave eye-witness accounts to scientist Helgi Eiríksson[98] (who later documented and reported these accounts) detailing how Fláajökull and

Heinabergsjökull met near Heinabergsjökull's eastern mountain of Geitakinn. This story of the ice meeting near Geitakinn was repeated to me by several semi-retired farmers in their eighties and nineties.

But many other people argue that the two glaciers did *not* meet, and a local scientist named Markús recently assessed the large area of land near to Geitakinn and discovered identifiable geomorphological features. As such, he concluded that Fláajökull and Heinabergsjökull never merged. Markús talked me through the issue: "That's the problem. Did it touch, or did it not touch? What I found out is that it was impossible for it to do so [touch]. The reason for it is you can actually see the end moraine, the assumed end moraine, and the glacier has not damaged anything there. It should have damaged everything."

It should have cleared away all the sand in the sandbox.

I asked Markús about how many people in the community repeated stories about how the two glaciers merged together. Markús shook his head. "What I found was that there was the remnants of a house here, and in this grass and soil you can find a tephra layer from 1300 Öræfajökull eruption in 1362."

In Markús's words resides the heart of the issue: "remnants of a house." The reason Icelanders in this region debate whether Fláajökull and Heinabergsjökull met is because of what was located in the place where the ice supposedly merged, what the two fingers of ice slowly pinched. At one time, the farm Heinar sat stuck between the glaciers' two surging snouts.

The farm Heinar was either completely crushed by ice, or, at best stranded, pinched by an icy claw, by two enormous advancing glaciers. All access routes were cut off. Oral stories tell that the farm was abandoned in the later part of the nineteenth century, and all those who lived there scattered, folded into Hornafjörður's local society.

Again, try to imagine the terrible decision the people at Heinar had to make; how long they waited as the ice grew closer and closer, masticating pastureland and mountainsides as it approached. This was no flood that surged; this was days, weeks, months, and each day filled with the terrible crushing crunch of ice creaking forward. An advancing glacier is a horrifying orchestra.

The farm Haukafell, which was on the opposite side of Fláajökull, was destroyed *twice* by Fláajökull's advancing ice before it was finally abandoned completely. All the inhabitants, including farmer Vilhjalmur Gudmundsson, his wife and their children, fled from their home before the power of the ice.

At what point did the people choose to pack up the family and run? Did the people of Heinar wait too long? Were they able to safely get across the dangerous rivers, the surging ice? Over and over, the people of Mýrar faced the power of ice, and that power has seared the social imaginary of generations of people living in the area. People would recount such stories to me as if they happened yesterday.

This is because the story of Haukafell is also more broadly the story of the district of Mýrar. When local farmers—many of whom were in the region's retirement home when I chatted with them—talked about Haukafell, they talked also about how the entire district of Mýrar had to continually work together to fight off the ice, jökulhlaups, floods, and continual marshland encroachment.

Twice I visited Haukafell with locals. The first time, I hiked out to the site upon the recommendation of a farmer, Silja, an older woman with crisp chin-length white hair that I'd had coffee with the day previous. Born in Vík, Silja had lived in Mýrar since 1980. When I met her at her farmhouse, she recounted several different stories about living with glaciers, about her farm losing land when the Icelandic government initiated the 1998 Vatnaj*ö*kull National Park boundaries, and the events of Haukafell. But first, she told me stories of her early impressions of living in the area. "It is really easy to see how the nature has drawn the story into the landscape. . . . It is one of the things that is most impressive for me to see here. Because you see so easily how the glaciers were two kilometers closer to here [Silja's farmhouse] than they are now."

As she spoke, Silja pointed out the window to the clearly visible moraines and waterlines in the mountains that suggested where the ice used to be.

"I did not grow up here," she said, but she had heard all about how "the water was running all over the land, floating all over." The woman who lived at the farm before Silja told her stories of how the water level would rise all the way up to where the stables were today. "She would tell us about Haukafell, about how children from that farm could be in their house and throw stones at the glacier . . ."

Silja told me that in modern times, all that remained of Haukafell were clear rectangular depressions in the ground where the turf buildings once stood. Most local accounts suggest Haukafell was abandoned in 1937, and that it was the *second* site the inhabitants had used before they gave up entirely on the farm. The first farm was further west, and records indicate people lived there starting around 1658-1697. But that was right during the Little Ice Age cooling, and over the decades, Fláajökull slowly advanced and retreated—a devastating pattern where it grew bigger, retreated a little, but then advanced even more. By the end of the nineteenth century, the glacier nearly blocked the entrance to the eastern valley Kolgrafardalur.

Fláajökull advanced so much that it crushed and obliterated the original Haukafell farm site around 1900. Then the family moved to a higher elevation slightly to the west, salvaging what they could and building several new turf structures: a residence, cowshed, storage houses, sheepcote, stables, and several miscellaneous sheds.[77] But by 1937, Silja remembered, glacier floods from the blocked valley were devastating the family's pastures, paths, and the physical land on which the primary structures stood. When the family finally abandoned the farm the second time, they deconstructed the valuable timber of the buildings to ease their welcome elsewhere, and like the family at Heinar, they too dispersed throughout Hornafjörður's society.

I hiked out to Haukafell after hearing Silja's stories. The site was at the end of a long dirt road, slightly elevated on a hill over Mýrar's outwash plain. Fláajökull loomed just over the mountain Fláfjall. Even with the minimal elevation gain, turning around to look away from the ice and back out over the sandur was stunning—the plain was dark, poorly vegetated, and riddled with palaeochannels, glaciofluvial terraces, moraines, roads, sparse settlements, and then the North Atlantic Ocean. It was extremely clear standing there how vulnerable the entire region was.

From my vantage point, I could see that any jökulhlaup wave of ice-water-slurry released by Fláajökull would have had no barriers to slow it from barreling over the land (or a house, or a person, or a child) as it headed single-mindedly to the ocean. Out to sea and gone. Straight south of where I stood, the next landmass south of Haukafell was Antarctica.

I visited Haukafell a second time when I walked out for an excursion of Fláajökull's forelands with the farmer Smári and his son Kristófer. Kristófer was in his early 30s, and he'd worked for the last five years at the school in Höfn. Kristófer's grandfather had purchased the family farm in Mýrar in 1952. At that time, a fair amount of Fláajökull was included in the sale.

Directly in front of us as the three of us walked was Fláajökull, and to the right was Haukafell. Kristófer talked about how, for locals, Haukafell was a special place. He said that Haukafell reminded locals what the people who used to live in the region went through year after year. Today, he pointed out, people are gradually replanting the area around Haukafell with trees. "In the eyes of the local people, this area right here is quite holy for people to just go and relax you know. We don't send tourists there. . . . We want this for our own."

Smári listened to his son talk, and between waves of stories, he told his own, about how other farmers had to incrementally move their farms south as the ice advanced, and later, as the ice stopped advancing but glacier floods increased.

"It was bad then, but then, when I was here and growing up it was less bad. We worked hard to make it better. Look around here today. We are spreading out and living where we could not have lived before."

Smári told me story after story of how the people of Mýrar worked together to push back against Fláajökull—stories he was told by his father, Leifur, who grew up in Mýrar. Later, I visited Leifur, who was in his 90s and lived in Höfn's retirement home. Like many of the older Icelanders I came to know, he was fit and spry. I had no idea he was in his nineties until he told me.

Leifur repeated to me the same stories of Mýrar that other farmers and older retired people living in the area recited. The stories of Mýrar, and the power of Fláajökull, and the stories of why people here were not particularly upset that the glaciers were disappearing—these were part of the cultural fabric of this place and it seemed everyone here knew them. Then he told me the story of the battle of the Hólmsá—Fláajökull's primary outlet river.

In 1936, around the time the family at Haukafell was forced to abandon their farm the second time, the few farmers left in Mýrar gathered together and wrote a letter to the Alþingi (the Icelandic parliament), testifying that if nothing was done about the glacier rivers (especially the local river Hólmsá) and the catastrophic glacier floods, then the district of Mýrar would be ruined. There were too few farmers each year, and the glaciers were too powerful.

The problem's epicenter was Fláajökull. Fláajökull's terminus fluctuated quite a bit in the 1930s, and subsequently its outlet river, Hólmsá, fluctuated as well. The river began to veer eastwards, gradually taking over another smaller outlet river called Hleypilækur, and eventually, another river further east, the Djúpá. The three rivers Hólmsá, Hleypilækur, and Djúpá covered an expanse of over half the entire farm district of Mýrar, and the floods and jökulhlaups that surged over and over down those bursting waterways destroyed phenomenal swathes of agricultural land, roadways, and bridges, and marooned or threatened over twenty farms.

The government's answer to the Mýrar farmers' letter of distress arrived in 1937 in the form of a single engineer bearing shovels and aerial photographs of the district. The engineer met with the farmers, surveyed the area, and determined the best course of action was to force the river Hólmsá back into its original channel.

All the farmers had to do to fight the glacier was move a river.

So for two months in 1937, most of the area's local men got to work. Mechanized bulldozers did not arrive in Mýrar until after 1955, so the farmers, armed with shovels and picks, cut *by hand* through one of Fláajökull's crested moraines, dug a new channel to redirect the river, and used wagons loaded with gravel to build an enormous earthen dam to divert the Hólmsá.

Leifur richly narrated the story of moving the Hólmsá river to me for over an hour. He paused to stare at his hands as he recalled the enormous amount of work of shoveling sand for two solid months. Nearly eighty years after moving a glacial river, Leifur still saw the shovel in his hands, still gripped the wooden handle—the thing between him and losing his farm, losing his community to the glacier and its floods.

When I asked Leifur what he thought about glaciers then, he just looked at me. I thought at the time that he didn't understand, and I asked his grandson Kristófer to repeat my question. Leifur waved at me—he'd understood the question, but he didn't think it was important.

After a few minutes, he answered. "We always thought about glaciers."

. N .
V ✣ A
· S ·

IT IS USEFUL, for just a moment, to dive a little deeper and ponder what is meant when we talk about power. About glacier power. About how for hundreds of years Icelanders had to make decisions to go or stay based on glacier conditions—the same decisions people everywhere at various times make based on their environmental conditions. About how today some Icelanders in Hornafjörður (and many people worldwide) express their powerlessness before glaciers—but this time in the context of their own inabilities to stop or slow the process of glacier retreat.[85]

We know that power is more than one thing's ability to act upon another, and that power moves through and inhabits all spaces of modern life.[99-101] But our specific definitions of power—and where and how it is located and enacted—changes as scholars worldwide assess and re-assess theorizations based on a continually fluctuating world.

The idea that you or I, or nations, or corporations, possess power is seemingly a given. No one would blink at the idea that the president of a country, or a large online retail giant, possesses power. As such, what many scholars focus on today is not so much the possession of power as how power is possessed, and how power is constructed, deployed, enacted, performed, assembled.[102]

The focus point here is "assemble." Assemblage and power go hand in hand.

A bit of back history: in the 1980s, French scholars Gilles Deleuze and Félix Guattari put forth their idea that parts within a whole are not fixed nor stable—rather that all parts are fluid based on how they continually relate and re-relate to each other within a greater constellation of the whole.[103] There is much more to it than this, but the important part is that all things within any given constellation are fluid. They called their idea assemblage theory. Now, the term assemblage struggles when translated into English—in French, the original term, "agencement," associates with ongoing, uncertain, unfolding processes and is simultaneously a verb and a noun. In English, "assemblage" becomes fixed, final—the antonym of what the scholars were trying to describe.

What scholars are interested in today is how certain arrangements of elements within a whole create and enact different power. How certain assemblages centralize or decentralize power, or how specific assemblages within specific spaces—such as institutions, relationships, or agencies— construct, produce, or deploy power.

For example, take an arrangement of three things. You, me, and a puppy. What power relationships might be there? What would happen if the puppy were switched for a weapon? Might the power dynamics shift? What if the weapon was held by me, and not you? How things are assembled together matters, and produce different power arrangements, but those power arrangements are not static. I might drop that weapon, and you might pick it up, and now the dynamics have shifted. Power can be a product of how elements orbit each other within the whole.

Jumping back to glaciers: it is not hard to envision humanity's ability to enact power upon ice—most recently through the forcing of global glacier recession via specific arrangements of elements including anthropogenic greenhouse gas emissions and rising temperatures. But said aloud the other way—glacier power enacted upon people—might give pause. However, glacier power enacted upon people within the context (read: the assemblage) of Mýrar, Heinar, Haukafell, Leifur, glacier surges, and jökulhlaups—suddenly the idea of glacier power doesn't seem so far-fetched.

Take this a step further. Power, even centralized, is rarely sourced from a single component. Glacier power does not come from a single point within a glacier. It is assembled in context of geography, scale, time, people. It is assembled within continually shifting constellations.

Some might argue that it makes sense that (broadly) people possess power—people are, after all, conscious living beings. But minerals, or matter, or inanimate objects, or environmental phenomena such as glaciers—these things cannot possess power.

In response to this, over the last several decades, scholars have shown power within more than human things (scholars often use the term *nonhuman*, but as I find it places primacy at humanity's feet I replace its use with the equalizing term *more than human*). Prominent examples include assemblages of dams, malarial parasites, synthetic chemicals, famine, and war in Egypt;[104] tidewater rice cultivation, slaves, and the Georgian landscape;[105, 106] and mushrooms, forests, pickers, and capitalism.[107] In this research, more than human things are shown to possess power independently and in conjunction with their relations with people and other elements.

Likely one of the most prominent scholars in this field is political theorist Jane Bennett, who conceives of a "thing-power" within more than human elements.[108] "I want to give voice to a *less specifically human* kind of materiality, to make manifest what I call 'thing-power,'" Bennett writes, "Flower Power, Black Power, Girl Power. *Thing Power*: the curious ability of inanimate things to animate, to act, to produce effects dramatic and subtle."[108]

What Bennett articulates in her thing-power is the importance of thinking about how the "material body always resides within some assemblage or other, and its thing-power *is a function of that grouping*. A thing has power by virtue of its operating *in conjunction* with other things."[109] Power is not enacted alone; a person does not have power in a vacuum, neither does a glacier. It is the assemblage of things together—regardless if they are human or more than human—and how those things shift within the whole that produces and directs power.

Thinking about some thing's power—in this case, glacier power—is critical. After Bennett, it emphasizes the shared connections amongst things (people and ice), the reciprocity of influence (glacier surges and farmers, anthropogenic greenhouse gases and melting ice), and how all things (*any* thing, be it person, glacier, mineral, vegetable) act in relation to each other without a permanent centralized pivot (humanity is not the center of the universe).[109]

Several hundred years ago, glacier power might have been readily visible in the form of glacier surges and jökulhlaups. Researchers Krista McKinzey, Rannveig Olafsdóttir, and Andrew Dugmore recognized the power glaciers possess when they wrote, "[t]he people of southeast Iceland have developed strategies to contend with both climatic variability and accumulated climate change as expressed through glacier fluctuations throughout their history."[77]

The expression of glacier fluctuation. This could be read: locals recognized and learned how to negotiate glacier power. The people of Mýrar developed strategies over centuries for engaging with glacier power *because* they lived with powerful glaciers. Mýrar residents placed their homes atop rocky outcroppings safe from glacier-powered floods and jökulhlaups. Those that did not, or placed their homes too low, or in the wrong place, risked their livelihoods like those at Heinar and Haukafell. Glaciers enacted power upon the settlement decisions of Icelanders.

As the ice receded and melted, rivers grew with greater meltwater discharge and people learned how to move the rivers. Today, glacier power shifts with new (read: warmer) environmental conditions: rivers have settled back down into their banks and are covered with disposable bridges; new homes have sprouted up on the sandur, down off rocky pedestals. Some locals point to how the glaciers seem weaker, have less power.

But even then, still, the legacy of glacier power remains. Höfn, the central village in Hornafjörður, is built on the farthest tip of a narrow spit of land stretching out south into the ocean. It is as far away from the area's glaciers as one can conceivably build a village and still be on dry land.

One young local man, Vatnar, commented on Höfn's geographic relationship with ice: "You can also imagine with how much it shrunk since maybe 1,000 years ago, how much the glaciers have shrunk. You can imagine how they were 1,500, 1,400 years ago, when people were settling here, just how large and intimating, just imposing across the country." Vatnar placed his left hand on the table between us and circled it with his right hand, gesturing that his hand was the island. "The glacier was so big, so that was like a fundamental part of okay, just living here at the beach, because all the towns are just at the beach, because the glacier was just fucking there, you couldn't go anywhere and the glacier was just there."

Human habitation throughout the region was in part influenced by ice, directly—Höfn is a space glaciers cannot reach—and indirectly—the sediments coming from the glaciers fill the lagoon, directing boat traffic to dock on the far tip of the spit where Höfn is located. For hundreds of kilometers along the south coast, Höfn is the only harbor.

Höfn in itself, however, is not unique. As Vatnar described, most settlements in Iceland ring the coast, usually corresponding with good harbors. But to be clear, Vatnar's point is both one of mechanics—the glacier does not reach Höfn—and one of perception. People in this region live "at the beach" in part also because of the perception of powerful glaciers.

Beyond habitation, glacier power is discernible as glaciers continually challenge human mobility, access, and borders. What do you do when the property line for your farm, your national park, or the barrier separating your flocks of sheep, dissolves? Or grows?

Glaciers have long acted as borders throughout Hornafjörður, restricting, fencing, protecting, and delineating. But borders (fences, walls, etc.) are tricky things. Like glaciers, they are physical things that most certainly may *stop* or *block* some things. But borders delineate only if something *needs* delineation. Which means, a glacier is not a border alone. Borders, fences, walls are as Costica Bradatan writes, "things of the mind."[110] Walls, from *Game of Throne's* Wall, to Haiden's Wall, the Berlin Wall, the Glacier-as-Wall, all possess power rooted in a determination of whatever needs to be bordered (or fenced, or walled, etc.) and just how effective the glacier is within that determination.

For example, glaciers have long powered separation amongst animals in Hornafjörður. In Mýrar, like the rest of Iceland, animals were rarely fenced. Flocks of sheep were kept separated by glaciers. But today, as young farmer Gunnlaugur reported, "All the sheep here are from six, seven farms, and if the glacier retreats that much"—gesturing with his hands spread wide—"then the sheep can go all around. The effort in getting them then probably won't pay up as much to have livestock wild."

What Gunnlaugur worries about is that the glaciers have always acted as fences, and as they dissolve, farmers won't be able to afford the needed physical fencing, and as such, farms might not have sheep in the future. Then, a glacier won't be a fence. A fence acts only *as a* fence when something needs fencing; only then does the fence have the power *to fence* when it participates in fencing.

The sheep also must be kept separate. Other farmers spoke of a secondary concern to the loss of their glaciers-as-fences: disease. As the animals range further, they will potentially be exposed to more diseases from other sheep. Currently, livestock on the island—be it sheep, cows, goats, or other—are kept regionally isolated in case a disease infects an entire flock. In Hornafjörður, glaciers empower boundaries of separation between animals and disease.

In years past, Icelanders moved their sheep from pasture to pasture via glacial rivers, through valleys and marshes, or up and over glaciers. But if the glacier or river was too dangerous, the grazing was cut off and farmers had to find other ways to feed their animals.

Árvök, a woman in her fifties who chatted me up in Höfn's grocery store one mid-November day, reminisced: "My grandfather used to tell me stories about going up onto the glacier to look for sheep. That was part of their life. They had to go up to the glacier." Farmers had to go up on the glaciers to get access to isolated ice-free pastures typically near nunataks, or look for sheep, or other tasks.

At Höfn's retirement home, Leifur and other older farmers told me about trying to access pastureland in the 1930s-1950s, but by "then the glacier was so tall [I] had to get to the glacier by walking up the mountain, and getting onto the ice by going on at the side. You couldn't go straight onto the ice."

Normally, a farmer would have approached a glacier at the terminus, where the glacier was the thinnest, to get onto the ice. But in Leifur's story, the glacier grew so big it could not be approached head-on, and the farmers had to go up the mountains and cross the ice from the side.

Today, the farmers told me, the opposite occurs: many farmers watch for new pastures to open for access as the ice melts. Leifur and others told me stories passed generation to generation about rich pastureland in the middle of the glacier Hoffellsjökull around the mountain Nýjunúpar. "From 1600s, when you could go from the flatlands, walking, to this place, Nýjunúpar. You couldn't for a long time, but now, there are only two glaciers, I noted, in the way, and it is almost possible again." The farmers noted that soon they would be able to access this pasture as the ice decreases. The ice had covered the pasture since about the 1600s, but still the farmers remembered the richness of the grass there and they waited, decade after decade for hundreds of years for the land to be ice-free.

Glaciers also act as mental borders. Silja spoke about losing part of her farm's land claim on Fláajökull when Vatnajökull National Park's boundaries were initiated—stories echoed by other local landowners. To her, the glacier was once part of her family property, but now when she sees Fláajökull she sees a boundary; the glacier is a Vatnajökull National Park boundary and the distinction has changed her relations with the ice. Vatnajökull National Park was formerly established as a national park in 2008, and its boundaries (where possible) were determined by the 1998 existing white ice termini of Vatnajökull and the outlets. What this means is that for Silja, the farthest extent of Fláajökull in 1998 determined where her farm property line was newly established.

Walking the landscape now, there are few, if any, distinctions that demonstrate a change of ownership, a boundary, fences or walls near Fláajökull. But, whereas people once looked at the glacier and saw their own property, they now see a government-created boundary that determines, for example, land use practices. The border is largely in the mind.

During a conversation in Nýheimar, two young men from Höfn discussed the differences in the landscape as the ice disappears north and west of Höfn. One man, Sigurfinnur, a fidgety fellow with long lanky hair, observed, "I can't imagine being able to go into that direction [north]! Cause it's just ice, and we can't go there now. Like in fifty or a hundred years, there will be just mountains, and we can go over them, and that's just so strange!"

His friend, Heiðar, responded, "And we could build a new town! I can't imagine how it would be though."

Sigurfinnur immediately retorted: "I think that's the strangest thing. We have only two directions now, we have south and we have east. And then we will have a third direction. The middle, that's mind-blowing."

For the two young men, at present they could only range towards Reykjavík, or east. North and west were impassable. The British writer Robert MacFarlane, writing about mountains, once observed:

> What we call a mountain is thus in fact a collaboration of the physical forms of the world with the imagination of humans- a mountain of the mind. And the way people behave towards mountains has little or nothing to do with the actual objects of rock and ice themselves… Mountains-like deserts, polar tundra, deep oceans, jungles and all the other wild landscapes that we have romanticized into being- are simply there, and there they remain, their physical structures rearranged gradually over time by the forces of geology and weather, but continuing to exist over and beyond human perceptions of them. But they are also the products of human perception; they have been *imagined* into existence down the centuries."[III]

MacFarlane uses the term "collaboration" where I would apply "assemblage," but the distinction separating mountains *just there* and humans *just here* is not quite on point. In his iteration, MacFarlane argues that mountains—vis-à-vis glaciers—exist in isolation until human perceptions imagine them into this world. This articulation is rather one-directional, and does not in the end support MacFarlane's label of collaboration. Collaboration implies multi-directional input, but in his articulation, human primacy dictates the type of assemblage a mountain may be. But along the southeastern coast of Iceland, glaciers and people influence each other, and glacier power influences human habitation and settlement patterns, mobility and access, borders—for people, parks, animals, plants, disease—that restrict, fence, and protect, and how people perceive space.

. N .
V ✦ A
· S ·

IN LATE APRIL, on a rainy bright evening in Höfn, I had dinner with two locals, Dadda and Dagfinnur. Dagfinnur was born and raised in Höfn while his wife, Dadda, grew up outside of Reykjavík. They'd recently moved back to Höfn after living for a couple of years in the capital region. We'd become friends over the last year, and I was glad to spend a quiet evening with them.

As we ate dinner, their one-year-old boy playfully demolished the stylish Ikea furniture and squirmed around our feet. Conversation slipped around the happenings in the village, including a car accident involving a light pole and a young driver, spiders suddenly coming out of the thawing ground, and two new migrating ducks arriving in the lagoon.

After dinner, Dadda served dessert in the living room on the coffee table eye-level with the toddler, who, mid-attempt to grab the sweets, was intercepted by his father. Conversation carried forth as if Dagfinnur did not have a squirming child upside-down on his lap. We began talking of ice.

Dagfinnur said to me, "I can't decide if I think the glaciers retreating is a good thing, or, I don't want them to grow again, right? Not with my child, this child." He shook his son for emphasis and the boy squealed joyfully. "Those floods! I guess I don't know, they grow or they retreat, if that's bad."

I asked Dagfinnur: "So they should just stay as is?" and he immediately replied: "Já. [Yes]. They should just be there. It was so bad before." Dadda voiced her agreement from across the room. It was so bad before—all those farms destroyed and people killed.

But now, they both observed, it was getting so much better. But they didn't want it to get too much "better." Neither of them, to use the words of the man at the opening of this chapter, wanted the glacier to lose *all* its teeth. Maybe just a few teeth, to make the bite less deadly.

Many Icelanders expressed similar conflicting views to me, listing simultaneously the positives and negatives of glacier recession. Locals did not like living next to powerful glaciers that threatened lives and property, but neither did they want glaciers to weaken and disappear.

When we look at the currents of power that course amongst people and ice, the waxing and waning of power as ice grows and shrinks, what comes into focus is the profound influence each holds upon the other. It is this influence that suggests that glacier change is never simple. That people can recall the ordinary horrors of times before and feel immense conflict over what they see happening before them today.

While the next chapter delves in more detail into local perceptions of glacier change, here it is enough to listen to Leifur. He told me that "when the glacier was melting, it was creating a sense of security in Mýrar that they [people of Mýrar] didn't have before. The glacier's going was something that allowed the people to thrive."

Glaciers, for centuries, enacted the power to destroy homes and livelihoods, to push people to the brink. To understand glacier power is to position glaciers beyond reductionist and static narratives, to recognize the potency of glacier power emergent from assemblages with people and the more than human. The stories and memories of powerful glaciers are alive in this region today, even as the power of glaciers themselves has weakened. At present, in Mýrar, Fláajökull has receded well over a mile and a half since 1890. The ice loses power and people gain power—but it is not an easy power to hold.

Imagine this was you. You'd lived with powerful glaciers your whole life, and you grew up steeped in stories of immense destruction. And today, before your eyes, those same glaciers shrink away. What would you do? Go or stay?

Glacier change is never simple. No change is simple.

CHAPTER FIVE

≈ ≈

"I have come to see climate change in an entirely new light: not as a media battle of science versus vested interests or truth versus fiction, but as the ultimate challenge to our ability to make sense of the world around us. More than any other issue it exposes the deepest workings of our minds, and shows our extraordinary and innate talent for seeing only what we want to see and disregarding what we prefer not to know."
GEORGE MARSHALL, 2015[112]

"Well, when you talk about global warming to Icelanders, they just look up. Really? Where? So it is getting warmer everywhere else and we are getting all the cold up here. So we are seeing snow all over, and we seeing, for us, we feel like the glaciers are getting bigger because it is coming so much snow is coming and so, so we are like, not really worried about them going away, or melting."
LOVÍSA, 2016

SEVERAL YEARS AGO, I was giving a lecture on glaciers in Reykjavík to a mixed audience of international physical and social scientists. While I spent a fair amount of time discussing Iceland's glaciers, I also spoke about many glaciers outside of Iceland, and how overall most glaciers planet-wide were in retreat.

At the end of the lecture, just before I was able to complete my concluding "thank you," a hand shot up in the audience, followed instantly by the entire standing body of an American man in his late sixties. He gestured impatiently at the coordinator, who was passing around hand-held microphones for audience member questions.

"What about Hubbard," he accused me, forgoing academic niceties of introducing himself. "What about Perito Moreno? Or, anything in the Karakoram?"

I nodded. I've been asked these question many times.

"I think it is ridiculous for you to stand up there and not mention all the glaciers that are advancing!" The man continued, aggressively cataloging specific glaciers worldwide that were either increasing in size or not decreasing in length. He concluded his question with the statement, "global warming is not making glaciers smaller, and I haven't seen any real evidence to suggest that it is even happening. I don't believe you."

Periodically throughout my career, I've encountered people who ask me questions about advancing/growing glaciers (usually much more politely) and what I explain is that glaciers have immense diversity and geography. Glaciers respond in multitudes of complex ways to local, regional, and global environmental dynamics, and no two glaciers are ever the same. Some glaciers—for reasons still not entirely clear—are indeed growing larger today, but the vast majority of glaciers are decreasing in size.

I explained these glacier nuances to the still-standing American man.

He'd chosen some of the most famous examples of advancing glaciers. The Hubbard Glacier in Alaska is a tidewater glacier responding to increased rates of precipitation in the Gulf of Alaska and is subsequently increasing in size.

The Perito Moreno is one of two well-known glaciers in Patagonia—the other is the Bruggen Glacier—that are advancing. Most glaciers in Patagonia are in retreat, but thus far, beyond speculations about specific micro-climates around each glacier, scientists have not yet figure out exactly why these two glaciers are growing as all the glaciers around them are decreasing.

There have also been significant rates of increased snowfall in the Karakoram mountains, leading to somewhat of a feeding frenzy amongst the glaciers there with many noticeably increasing. Again, local dynamics.

The event coordinator hovered near the man, visibly itching to remove the microphone from his hands. But he held fast, and as I concluded my response, he shook his head negatively. "But those glaciers are advancing!" he responded.

I nodded again, agreed with him, and moved to the next person, the next question.

The Hubbard, the Perito, and the handful of other glaciers in states of growth today stand as anomalies, but I have found that they register to some people as strong evidence that global trends of ice loss are not occurring, or not occurring at the rates scientists predict. There are over 400,000 glaciers[35] located worldwide, from tiny glacierets to rock glaciers to ice fields and ice sheets, and unfortunately, almost all of those ice bodies are in some way decreasing. But continuously, some people overlook the vast number of glaciers retreating and point out the few growing larger. These few increasing glaciers act almost as icy chimeras, passed from person to person in confirmation of wishful fictions and futures full of glaciers.

Iceland has its own "chimera glacier;" Heinabergsjökull, on the southeastern coast, is a thin outlet glacier between Fláajökull and Skálafellsjökull. From Höfn, looking west, low-lying Heinabergsjökull is easily obscured when a slight sandstorm kicks up or a marine layer settles in. Seen from the air, Heinabergsjökull is not unlike a long and flat finger pointing at the sea. The face of the glacier's terminus rises only 30-60 feet above the lagoon's water level.

Steep valley walls cradle the elongated body of the glacier, with the bare rock mountains Hafrafell (3,254 feet) and Geitakinn (2,358 feet) standing as gateposts on either side, the land between them long ground away by ice and deposited into the North Atlantic Ocean. Whereas the other five glaciers featured in this book cascade over Vatnajökull's encircling mountain ridges and sit high atop their carved beds, Heinabergsjökull eases down over a slight ridge off the ice cap and then stretches for fourteen straight slumped miles in a long, flat trough with a small over-deepened basin full of glacial meltwater at the end.

Heinabergsjökull's proglacial lake, Heinabergslón, is colored milk-chocolate by high glaciofluvial sediment loads and crowded with circulating icebergs often the size of small houses. Heinabergslón and the terminus of the glacier itself are no longer visible from Höfn; actually, they are not visible until visitors physically reach the car park after driving thirty minutes down a corrugated dirt track off the Ring Road.

Heinabergsjökull is receding—all glaciers on the southeastern coast are in recession—but the glacier's recession is less visible than others such as Hoffellsjökull or Fláajökull. This has led to Heinabergsjökull achieving local "chimera" status.

Whereas other local glaciers are losing ice mass in a variety of ways, including along the entire length of the glacier (making the ice appear thinner each year) and from the position of the terminus (making the glacier appear to colloquially "retreat"), Heinabergsjökull's terminus position has shifted very little over the last several decades. To be clear, Heinabergsjökull is actively receding due to increasing temperatures,[40, 113] but the *process of recession*—the way recession looks for this particular glacier—makes it harder for people standing in the car park to detect.

Language fails here. Glaciers technically recede, meaning they lose ice volume in response to a negative mass balance. When a glacier is in recession, both the length and thickness of the glacier body can decrease. It does not, per se, retreat—as in, move backward. However, popular discourse continues to use the term "retreat" to describe the process of recession, and people subsequently expect that glacier recession looks like retreat—with the terminus of a glacier appreciably moving back from a specific point.

What is happening to Heinabergsjökull is tricky. Of all of Vatnajökull's outlet glaciers, Heinabergsjökull is one of three glaciers experiencing the highest rates of downwasting, the reduction in thickness in the ablation area of the glacier.[34] Once the glacier flows off Vatnajökull and drops down into the valley, fourteen miles of its long thick body are all within the glacier's ablation zone. The ablation zone on a glacier is the area where ablation exceeds accumulation, where the glacier loses ice mass through melting, erosion, sublimation, or calving. Critically, the *entire length* of Heinabergsjökull is ablating, all fourteen miles.

Geography is significant to what is happening to Heinabergsjökull. Like all the glaciers in the area, over the last century Heinabergsjökull has advanced and receded repeatedly, oscillating back and forth. The constant see-sawing back and forth has been like sandpaper on cedar wood, and the glacier has gouged deeper and deeper into its mountainous trough. And as Heinabergsjökull has dug (sunk) deeper, it has created two interesting effects that make the glacier quite different from its neighbors.

First, the proglacial lake at Heinabergsjökull's terminus is increasing in size *not out but in*, backfilling over and under the body of the glacier itself. While the glacial lagoon has an outlet—the river Kolgríma—the river has a low flow rate because the lake itself is below sea level and the glacier dredged its trough much deeper at its northwestern end than on the opposite end where the glacier presently terminates. As such, as Heinabergsjökull increasingly melts due to warmer temperatures, the meltwater coming off the glacier cannot flow to the ocean; instead, it pools in the lake and backfills *into* the trough—trying to fill those deeper parts of the trough—and seeks to flow under the glacier body itself.

Second, the erosional power of meltwater flowing back under the glacier is quite strong. The warmer meltwater is slowly severing the body of the glacier from the bedrock. The ice that is lifted and separated from the bedrock floats like a tidewater glacier. This processed is called a "floating tongue," where the ice floats and stretches out over the surface of the water, increasing in length to maintain stability.

Crucially, it is all of these dynamics together—the deep trough, the recession, the meltwater backfilling, the floating tongue—that makes the terminus of Heinabergsjökull appear year after year as if it has not recessed at all. The glacier's terminus appears to be fixed in place—even as water has crept up its sides. Since 1890, the glacier has receded 1.8 miles from its maximum margin, but the recession has not been linear, and much of it has been disguised by the floating tongue.[40]

The take away is this: few of the mechanical subtleties of Heinabergsjökull are legible from the glacier's edge, or from the glacier's parking lot, or from the village of Höfn. Without specialized knowledge of glaciology, a person could reasonably assume that while all the other outlet glaciers in Hornafjörður were in recession, Heinabergsjökull was not. It might even be reasonable for someone to look at Heinabergsjökull and think that the glacier might be like the Perito Moreno or Hubbard. Might be growing, advancing. Not melting.

As such, Heinabergsjökull has achieved locally "chimera glacier" status.

In Höfn, some individuals I spoke with articulated opinions that glaciers on the southeastern coast were not receding. Others, like the man who questioned me in Reykjavík, argued that glaciers were growing. Some people told me that some glaciers were receding, but that other glaciers were not. Others still yet observed that while some glaciers were receding, it was likely all glaciers would return in the next decade or so.

Teenagers, teachers, municipality officials, fishermen, farmers, older people, men, women—people from across social sectors expressed such views. And often, when people discussed their opinion during in-depth conversation, in passing at the grocery store, or in a variety of other venues—they referenced Heinabergsjökull. Heinabergsjökull acted as evidence for whatever viewpoint they professed about local ice behaviors. While they might not have been personally out to visit Heinabergsjökull recently (it was rare to see people—tourists or locals—*at* Heinabergsjökull), they still pointed out that the terminus of Heinabergsjökull had not moved in decades, and that, in fact, the glacier was not receding at all.

Some people in Hornafjörður told stories of Heinabergsjökull to substantiate narratives of glaciers and glacier change that were contrary to the current material behaviors of most of the region's glaciers. They saw glacier recession in other glaciers but comprehended Heinabergsjökull as a glacier that advanced, and thus told stories of *all* glaciers advancing.

Importantly, many people in Hornafjörður discussed local glacier change as a definitive sign of unprecedented glacier loss, and many others associated rapid glacier change with global climatic changes. But for this chapter, the focus is on people who did not, and on the chimeric plasticity of glaciers to verify multiple conflicting narratives all at once.

I'm not interested in determining if the stories some local people told were untrue, but rather, it is more interesting (and telling) to consider the incompleteness of such stories. It is illuminating to understand why a scientist would focus on just a handful of glaciers in the face of 400,000, or to understand why some people make sense of melting ice in particular ways.

British environmentalist George Marshall started the prologue of his 2015 book *Don't Even Think About It: Why Our Brains Are Wired to Ignore Climate Change* with the story of Felix Frankfurter, a Supreme Court judge and a Jewish man, who heard eyewitness testimony in 1942 about the genocide of Polish Jews in the Belzec concentration camp. Regarding the eyewitness, Frankfurter said: "I must be frank. I am unable to believe him. . . . I did not say this man is lying. I said I am unable to believe him. There is a difference."[112]

The American man in Reykjavík did not accuse me of lying. He just said he did not believe me.

. N .
V ✥ A
· S ·

IN HÖFN, glaciers are visually accessible and immediate. Day after day, it is difficult to not see glaciers in ordinary life as locals go about their lives, commuting, walking, shopping, working, socializing in greater Hornafjörður. Glaciers are *just there* outside kitchen windows and offices, through the windscreens of cars and tractors, unmediated from about every standing position in the village not blocked by a one or two story building. Images showing how glaciers used to look at various times are posted in various private homes and guesthouses, on the town's waterfront social path, inside Nýheimar, at the entrance to the single grocery store, and other social spaces. As such, it is generally quite difficult to *not* see glaciers and *not* see how much the ice has changed—receded—over the last decades.

The high visibility of glaciers and glacier change in this area, however, does not stop some people from *not seeing*, or, more commonly, repurposing what they do see into contrary stories of environmental change that essentially bypass or misinterpret the material reality of glacier change on the southeastern coast today.

Such stories, typically attached to Heinabergsjökull, were diverse and evolving, and articulated in varying forms by the young and old, the retired and the early career, men and women, people in prominent positions in the village and those who held less visible jobs. It was difficult to identify a commonality or pattern in the *people* who voiced these interpretations—they were sourced from diverse sectors of society. People in Hornafjörður, broadly speaking, generally have good levels of education (many with university degrees), strong sources of income, social equality, Internet accessibility, long lives, and good health.

To demonstrate these diverse views of glaciers, in the following sections I'll highlight five specific people and the stories they told. I've chosen to highlight individual stories for each section instead of clumping in multiple examples for two reasons. First, I believe individual stories are much easier to follow than multiple fragments. And second, I believe I can present the widest array of glacier stories in this format. A reminder that I have taken liberties to anonymize each person to protect their identity.

. N .
V — A
· S ·

IN THE MIDDLE OF FEBRUARY, I sat in a comfortable chair in Lovísa's living room in Höfn. I had a cup of coffee in my hands, a plate of chocolates nearby, and I was waiting for Lovísa to tuck her younger children into bed before she could sit with me and talk.

The house was quiet. Lovísa's husband was at work, and the only sound was the quiet murmuring from the television show her older children were watching.

Lovísa, a compact woman in her mid-forties with blonde hair, heavy makeup, and sharp movements, walked into the living room, and as she picked up the remaining laundry, we discussed the weather—it had been below freezing for over a week and the recent snowfall was heavy. I was late arriving to her home because my car had gotten stuck in the snow.

Lovísa had a prominent position in town; often when I spoke with others in the community, they attributed certain stories and facts back to her. Lovísa was well educated, and, like many professional Icelanders, went abroad for her advanced degree. She quickly moved me through her background, her qualifications for her current position, and what her career prospects entailed.

Eventually, we came to the reason I was there: glaciers.

I barely asked her a single question before she took off on a variety of subjects unrelated to glaciers. After about ten minutes, we circled back, and I asked her about how young people in the area expressed worry about getting educated and then being unable to return to Höfn for a non-tourism related career.

Lovísa shrugged her shoulders and responded: "Actually, it doesn't really matter what you learn, you can always find a job here. Because like engineers, doctors, nurses, healthcare people, people living here always need service. Lawyers, and já [yes], the fisheries, that's a growing industry and they need more people to work with the fish. They also need people like thinking of stuff, já, and building stuff, and fishing industry is getting better, very specialized, and very technical. So you always need more educated people to work there."

"Why isn't that message trickling down?" I asked.

"Because they don't know anyway, maybe they have worked there with the bosses, but they don't know what's going on around them," Lovísa replied, "what all the bosses are doing, all the chiefs." She paused, sipped her cup, then continued. "Regarding the tourists, já, I still think that they will come here, although they may have to hike this one hour to get to the glaciers. Global warming we are not really seeing that here," she laughed. "Well, when you talk about global warming to Icelanders, they just look up. Really? Where? It is getting warmer everywhere else and we are getting all the cold up here. We are seeing snow all over, and we feel like the glaciers are getting bigger because so much more snow is coming and so, so we are like not really worried about them going away, or melting."

"You're not worried?" I asked her.

Lovísa shrugged her shoulders again. "It is not like a scientific view. I don't know this, I haven't measured this, I haven't read any articles or anything like that, but it is just what we experience. And sometimes we maybe we see some old pictures and then we see okay, some glacier that is like way further out now then it was fifty years ago, so okay, the glaciers come and go, they are crawling farther farther back and forth."

Lovísa extends her hand out, gesturing to show how much a glacier has retreated. "That's the normal thing. There was an ice age, and then it got warmer, and maybe we are still in that period or maybe we are heading back to the ice age, I don't know." She pauses, laughs a little. I nod.

"It is very hard for somebody to convince an Icelander that global warming is real, because we are really not seeing it. It's getting colder but that's something we're very used to because it's come some years with cold summers and hot summers and more years with cold summer."

. N .
V ⊕ A
. S .

I WAS WITH MY FRIEND Ylfa at the defunct NI gas station in the farm district Nes because they were having a shoe sale. Once a gas station, but now a sometime shop, every time I drove past the building something else was being sold. Today, it was shoes.

Ylfa and I sorted through the boxes of shoes together and found one pair for me and two for her. To celebrate, we went to the other side of the gas station, which was operating as a cafe.

We bought coffees and waffles and joined two of Ylfa's twenty-something friends, Friðmundur and Zóphanías, at one of three tables in the room.

Friðmundur and Zóphanías introduced themselves, and told me they both worked in the fishing industry in Höfn. I introduced myself in return and told them I was researching local glaciers.

Both men were skinny, with brown hair, brown eyes, and similar builds, and I incorrectly assumed they were brothers until they explained how they were related to Ylfa through different local familial connections. The four of us drank coffee and talked about hunting geese and reindeer. As the conversation progressed, I asked them how fishing was going:

Friðmundur settled into the topic of fishing. "Everybody is here for the lobster. But Sigurður Ólafsson [a local fishing boat/company] is not getting as much lobster as he used to. Last year, in April they were going for lobster and they had to go further down south to find the lobster and then they weren't bringing it back here."

"Why is the lobster moving?" I asked

"I don't know. There is a physical change where the lobster is going?" Friðmundur responded.

Zóphanías jumped in. "They said it is the temperature of the water on the coast, it is changing," he said.

"Is the water getting warmer?" I asked.

Zóphanías shook his head. "No, colder. It is colder in Iceland now."

"It is colder in Iceland now?" I asked.

Zóphanías nodded again to affirm his statement. "There are the snows here now and last winter it was really really cold. You know, right?" he added. "The glaciers are growing with the snow?"

I raised my eyebrows. "Maybe? Are any of the glaciers here growing?"

Friðmundur pointed out the gas station window. "Heinabergsjökull is bigger this year."

"They said Heinabergsjökull was getting bigger. If Skinney moves, at least that!" Zóphanías laughed.

I was confused. "I'm sorry, I don't understand."

Friðmundur clarified. "It was colder and Heinabergsjökull is bigger this year." And if the fishing company Skinney moved out of town, which was always a topic amongst the villagers, at least they would still have an advancing glacier.

"Who told you Heinabergsjökull is getting bigger?" I asked.

"Every year FAS [the local secondary school] measures Heinabergsjökull and most of the glaciers here since I have been here. The glaciers go back and forth. It is natural they do that and Heinabergsjökull is the snow now—" Zóphanías was cut off by Friðmundur.

"—he is going to grow with all the snow," Friðmundur informed us.

More people arrived at the gas station seeking coffee, waffles, and shoes, and they came up to greet us. The conversation rapidly moved away from glaciers.

. N .
V ✦ A
· S ·

JARÞRÚÐUR AND I sat at an enormous round table inside a large empty room in a countryside schoolhouse in mid-March. We both held cups of coffee, and the sounds of animated conversations happening in the other room spilled through the closed door.

Jarþrúður was an older Icelandic woman who worked for the national park and had lived in the region her entire life. Extremely engaging, she sought me out to talk and asked for an interview to discuss local glaciers.

Lively and emphatic, Jarþrúður had long curly hair—a trait passed down to her five-plus grandchildren. She showed me pictures of each at the beginning of our interview. I showed her pictures of my husband and my cat, and we slowly became acquainted with one another.

As we progressed along into the conversation, I asked her what she thought was happening to the glaciers in the area.

"I think they have been retreating," Jarþrúður told me. "But this snow this winter makes us think they may not be in danger because all of the snow here must add something to the glacier. I think most people think it is a natural wave. Sometimes the glaciers build up, sometimes they get smaller. But I think most people here might believe that the glacier might be retreating, that is in a way natural. Because they were growing and growing and growing, so why shouldn't they retreating and retreating and retreating? Some time? I think that is what most people think. Of course, there are volcanic eruptions, and volcanic eruptions have a great impact on the glacier, and we can't control it, nor predict it. Of course, we know the people are, people in the world, we are using some fuel, we are always using fuel, and even if you are a very strong environmentalist, you are using a fuel . . . even if we know this may not be the best thing."

"It's a problem, though. We are fighting ourselves. I don't know about solutions—" I said before Jarþrúður jumped in.

"Perhaps this is not our task. We can notice a problem, and we can talk about it, and someone else finds the solution. That is, we shouldn't limit us to problems we can solve. Everybody here is a little sad as the glaciers they are retreating. For me, it was when Helgi Björnsson, Helgi Björnsson he is a, you know him, of course."

I nodded. Helgi Björnsson was one of Iceland's most famous glaciologists.

Jarþrúður continued. "When he [Helgi] came with a lecture here and he was telling us about how glaciers had been expanding bigger bigger bigger until 1860 or something, and then they started to retreat, that was a relief because when he said that, that was quite a new experience, because we had always heard that the glacier was retreating because of YOU. You are! Not us particularly, but the human beings are polluting, and destroying the glacier. So this natural wave expanding and decreasing, that he was explaining to us, it was a huge relief."

Jarþrúður drew her hand across her forehead, miming relief.

. N .
V ⊕ A
· S ·

IN EARLY APRIL, I was sitting at my desk in my office at Nýheimar staring out the window at the sunlight coloring the glaciers pink when a knock sounded on my door. Ari, a local man in his early forties, wanted to talk.

We had chatted informally over the last several months, but we had never had time for an actual sit-down interview. He was available now, he told me, and I invited him to grab a seat on the other chair in my office.

I cleared the books and papers away, dragged the chair out for him, and we started by talking about his family, kids, and his new tourism business. He was getting a significant amount of business, he told me—the rates of tourism were skyrocketing and it was a struggle for him to keep up. His bank account was brimming, he said, even though he saw his kids less.

Ari was a slow talker, thoughtful and careful with his words. He illustrated what he told me with frequent doodles on paper, emphasizing certain verbal points with blue ink dots tapped out slowly as he talked. After discussing his family and his business, we moved to glaciers.

I asked him, "A moment ago you said this whole area has a history with glaciers. Do you think it has a future with glaciers?"

"Yes, definitely. Especially if, well, how do you define future?" Ari responded.

"The near future," I replied.

"The near future. They are definitely getting smaller and smaller, so, but I actually believe that between now and 150 years,

I think there will be some changes, that might uh, I don't believe that glaciers will disappear in 150 years."

"In the future, not in the past?" I asked.

Ari nodded. "In the future. I think something will cause another ice age in between. So. Maybe it is through some catastrophe or something like that. But, I think I think so, but who knows?"

I inquired further. "Why do you think so?"

"I think, I think there will be some catastrophe in the world, that will change things around. And then the human they have to adapt to new conditions, circumstances, so, one thing could be more volcanic activity, and..." Ari trailed off.

"That would definitely do it."

"Yes, and one thing could be the gulf stream, and maybe we might see some changes in the future, maybe, after when we see more extreme changes in the weather, maybe something reduce the pollution, but I think that some natural causes might fix it before, but, uh, yes."

. N .
V ✦ A
· S ·

I HAD KNOWN HREFNA for several years through her work as a guide on the area's glaciers. She was in her twenties, tall, with long red hair braided into a thick rope that often stuck out from underneath her work helmet. She interacted with the area's glaciers daily for at least half of each year.

We sat down for an interview in early February in Hrefna's off-season when she was not guiding.

She had just returned from spending her summer wages traveling in South America, skiing, and having as much fun as she could dig up. She glowed as she recounted her adventures and listed all the great places she skied, and what was next on her list after another summer of work in Iceland. Her passion for the outdoors was evident throughout our conversation.

Hrefna told me about how she'd grown up in Höfn, but how many of the people in the area did not match her passion for the outdoors.

I asked her what she thought was happening to glaciers in the area.

She responded: "I think about it all the time, but if you read a lot, people have really really like different perspectives on it and actually what is causing them to melt, is that human? Or is it just actually the natural recycle of the glaciers? You know, they melt for some time, and then they grow again. So, I think for me every little thing about it I want to make my, I want to have some influence on it so I try to like recycle everything and think about how much I drive or, you know, everything. But still I am not, like, I am on both sides pretty much. I think there is, I feel like there is something that is just normal?"

"What do you think is happening to the ice? Because you said that you were on both sides of the fence? Can you explain that? What do you mean?" I asked her.

She explained in more detail. "So, I think because if you look they have been smaller around Settlement. And, there was a bit smaller glaciers in Iceland, and more grown land, so I am thinking, ok, maybe it is just uh, uh, like some scientists say that's actually what's happening, like recycle of glaciers, they get smaller for some time, and then at some point they are going to grow again. You know, but it is hard to tell, because now they say, for example, you see the snow that is in Höfn now, and they say like two years ago it was said that Iceland was going through some kind of cold climate change for few years, like it was going to colder winter and colder summer and then, the last two summer have been much colder than before and the winter has lasted longer and so on . . . So, for me, it's like okay, now when you have all this cold thing in Iceland are the glaciers gonna grow and in the highlands like Hofsjökull, last year, where it is not retreating, it is not getting smaller, which is the first time in a long time. It is really good, but then, on the other hand, it's hard to tell, because can you imagine, like the people in Iceland, or in Höfn, if we could really effect what is going on with the glaciers . . ."

"Do you think that the ice will come back?" I asked.

"Not that silly, but I am not believing that it is going to go away like all the way, like it is now. Do you see what I mean? If it is gonna keep going, I am gonna get concerned. I see both changes on Svínafellsjökull, like, it is melting, but it is still, but it is still coming down a lot, like some parts of the tongue have come 25 meters down, and sometimes you could see it has moved forwards although it is getting thinner, but it has moved forward a lot so, I don't know, so many different reactions to it" Hrefna finished.

TO BE CLEAR: glaciers in Hornafjörður *are* receding, including Heinabergsjökull, albeit with different rates.[1-3] Research in Iceland confirming glacier recession is *not* in scientific contention, and glaciology models predict that Icelandic glaciers will lose 25-35% of present volume over the next fifty years, largely due to global climatic changes.[3, 40, 41]

But in everyday life, the material reality of glacier change is not so clear—rather, it is narrated in a variety of contentious ways.

It is not useful to explore local stories of glacier change with the intent to prove or disprove them—that would actually tell us very little. It is likely unsurprising that people hold diverse and contrary views of glacier change—as people worldwide hold diverse and contrary views about most things.

It is much more interesting to focus on what contrary narratives tell us about how people interpret glacier change and make sense of their environments and landscapes. I use the term "interpret" here to draw attention to the hermeneutic nature of glacier change. Assumptions are often made that if we just explain the metrics of glacier change—for example, X glacier retreated one mile—there is no room for any other interpretation. What the preceding stories show us, however, is that there is always room for interpretation.

It is too simplistic to frame narratives of glacier change in various one dimensional believe/not believe, right/wrong, or skeptic-denier/believer binaries. Such approaches fail to allow for the complex ranges of individual and community interpretations and perceptions, different experiences and psychological responses, the influences of cultural histories, and other such variables.[114, 115] Essentially, such approaches overlook the *process* of narrative-making, of understanding how local knowledge is produced, authorized, and circulated across time and space.

So what is happening here? How do we make sense of contrary interpretations of glacier change?

Often times, when someone holds a counter opinion regarding environmental phenomena, they are accused of being ignorant or in denial or of not knowing. Think about how many times you've accused someone who did not agree with you of being ignorant, or in denial. I know I've done it many times.

Scholars across different fields have examined how people respond to other climate change-related information, specifically considering whether those people might be in states of ignorance or denial. This is an incredibly important area of research.

Is it possible that Lovísa, Friðmundur, Zóphanías, Jarþrúður, Ari, and Hrefna are ignorant and simply do not possess all the facts, or perhaps are in denial? Lovísa, Friðmundur, and Zóphanías suggested glaciers were growing, Jarþrúður had learned glacier change was natural, Ari knew *something* would make glaciers grow again, and Hrefna struggled to believe humanity could actually impact glaciers.

First off, igorance is a strong term with negative social connotations, and it is not my intention to label any individual ignorant. Rather, I am curious if people might be experiencing the emotional state of ignorance, the "absence of knowledge, probabilistic uncertainty, inaccuracy, irrelevance, and other sources of not knowing."[116]

Are some Icelanders actively pursuing *not* learning about ice, or unaware of glaciers and glacier change and subsequently incomplete in their stories of ice? Tangled up in issues of ignorance is this idea of "non-knowledge." Non-knowledge is where knowledge is purposefully not acquired, as opposed to the purposeful pursuit of knowledge.[117-119] For instance, you might not want to know anything about domestic animal slaughter. You might just want to eat your burger. So you on purpose do not acquire the knowledge of how most animal meat is processed before it ends up in a grocery store. In this case, you're ignorant about this topic.

However, none of the Icelanders I interviewed expressed their choice in not pursuing glacier knowledge; rather, they appeared satisfied in their personal levels of glaciological knowledge—and how they subsequently interpreted that glaciological knowledge.

So I wouldn't say these Icelanders had "non-knowledge" of glaciers. But what about ignorance more generally?

For decades, scholars have scrutinized how people respond to environmental changes, and how information about the environment is transmitted and received.[120-122] Often, even amidst an experience of profound transformation—such as a flood or a wildfire, etc.—some people mentally turn away from relevant environmental information and continue as before.

For example, people may place their homes in known fire or flood zones, or they might continue smoking, over-use water in drought areas, or fail to recycle—even as they are told such behaviors might negatively impact them or their community. One reason for this behavior might be ignorance, which is generally defined as a lack of proper education or information. And typically, when people are labeled ignorant, that label arrives with a recommendation on how to decrease any existing knowledge gap through various awareness programs and outreach campaigns.[122, 123] Implicit here is that by decreasing perceived knowledge deficits, people might rationally engage in pro-environmental or prosocial behaviors. Said differently, the idea undergirding that recommendation is that if people knew more about something, they would behave better towards it.

However, ignorance does not appear to explain what is happening in Iceland. Lovísa had spent quite a few family vacations *on* the area's glaciers; later in our conversation she told me about the extended ice-related coursework she completed over the last several years. Lovísa knew a prodigious amount about glacial ice and did not appear to have a knowledge deficit in relation to glaciers.

Perhaps Friðmundur and Zóphanías had a knowledge deficit. Neither of them appeared to know all that much about glaciers specifically—when I asked what they learned in school, they said the only thing that stayed with them was that Vatnajökull was the largest glacier in Europe.

When I asked them if other specific glaciers were growing, they repeated their facts about Heinabergsjökull, demonstrating a sample size of one out the approximately 270 individually named glaciers in the country. Such selectivity was akin to cherry-picking, suppressing evidence, or fallacy of selective attention—which are all *choices* a person makes about a topic and not necessarily suggestive of a person failing to possess the requisite knowledge.

Genuinely though, Lovísa, Friðmundur, and Zóphanías appeared unaware of the crucial reaction time between snow, glacier mass balance, and visible responses at the glacier's snout. A short reaction time is usually decadal, with longer reaction times sometimes taking centuries.[34] Snow experienced one year does not grow a glacier that year; in Iceland, even with relatively temperate glaciers, the reaction time can be upwards of one hundred years for snow to process into glacier ice and travel outwards to the terminus of the outlets.[1]

Jarþrúður, an environmentally-oriented woman who worked at the national park, was surrounded by interpretive displays, educational talks about glaciers and climate change, and a cadre of highly educated and communicative scientists, rangers, and others. She attended a talk by the country's leading glaciologist, Helgi Björnsson, and came away free of the burden of climate change-accelerated glacier change.

I met with Helgi Björnsson not long after meeting Jarþrúður to speak with him about his presentation. I was curious what exactly he said that evening, as it showed up in many interviews with the area's residents.

Helgi told me that "this happens all the time. . . . They often ask me, climate is changing, is there natural variability in the climate? And then I say well yes . . . but I need a 45 minutes lecture to explain this but if they just ask me I admit there are fluctuations, so then they misinterpret my shorter answer."

Helgi told me that he specifically addressed the link between climate change and glacier change in his lectures. It appears that Jarþrúður was educated about glaciers, but she emphasized certain aspects of Helgi's lecture over others and overlooked the abundant information available in her work environment.

Ari did not deny glaciers were changing—he was not ignorant of what was occurring and did not have a per se knowledge deficit of ice; rather, he selectively focused on a future event that would inevitably solve the present situation.

Hrefna, a person with high exposure to glaciers and a significant glaciological knowledge as required by her job, could not be labeled ignorant about glaciers—she taught glaciology to clients from around the world daily for half of each year.

Ignorance—specifically here a knowledge deficit surrounding glaciers—is clearly part of this story, but I do not think it is the entire story: none of the six Icelanders representing contrary narratives of glacier change could be reasonably labeled ignorant about local glaciers.

In other places in the world with high rates of glacier change, such as the Alps, Himalaya, and Cordillera Real, scholars found that local people were aware of glacier changes even as they possessed stories and different levels of glacier knowledge.[22, 124, 125] What this suggests is that it is not the *amount* or *specificity* of glacier-related facts or knowledge that equated relating current glacier change to future glacier disappearance. Frankly, it is questionable how much possessing "all the facts" about glaciers actually matters.

If all Icelanders possessed considerable glacier-related knowledge, could speak with ease about evapotranspiration, ice core stratigraphy, or mass balance budgets, would this impact their responses, their interpretations and narratives of glacier change? I suspect not. Measuring the amount or validity of a person's glacier-related knowledges does not explain contrary interpretations of change. Recent findings suggest neither a knowledge deficit nor ignorance is a principal indication in how a person receives knowledge; rather, it is more often the emotional and psychological state of the person that controls how environmental information is transmitted and received.[126, 127]

As such, I do not believe that production and distribution of contrary narratives in the area is based on some Icelanders not "knowing enough" about glaciers or being ignorant of ice. Neither explains what is happening here.

In this case, assigning ignorance as the primary factor for different readings of glacier change is simply a decoy. Such labels are low hanging fruit that distract from what is really happening.

. N .
V ✦ A
. S .

LET'S LOOK AT DENIAL. Perhaps denial is what is as play here, and explains Lovísa, Friðmundur, Zóphanías, Jarþrúður, Ari, and Hrefna's views.

Perhaps they are aware of the material realities of glacier change, but choose to deny or refute specific glacier-related information due to a complex set of emotional and psychological variables? I can relate to that.

Some studies have researched people and denial at different scales,[22, 26, 85] but most relevant to Höfn are studies researching at the community level—so beyond just one individual's response. Rather, these studies have looked at the entire community.

For example, from 2000-2001, sociologist Kari Norgaard explored the social denial of climate change in a rural community in Norway quite similar to Höfn. To make sense of her research, she examined her data using sociologist Stanley Cohen's theorizations of denial including literal (absolute dismissal of information), interpretive (reinterpretation of information), and implicatory (the psychological, political, or morals implications of specific information are diminished).[128]

Norgaard focused mostly on implicatory denial in her research, arguing that successful denial is socially organized and predicated on distancing climate change. Norgaard demonstrated that denial was not absolute—nor an individual's active choice—but rather a slippery and ambiguous process existing in the spaces between all-out refusal and inability to see, what Norgaard termed the zone between "knowing and not knowing."[128] Norgaard argued that her research participants evaded climate change knowledge without fully committing to that evasion; rather, they possessed a partial knowledge that was enough to answer for or cast doubt upon the veracity of climate change as a whole. In essence, Norgaard found willful incomplete knowledge.

As such, perhaps more than ignorance, denial might work to explain Lovísa, Friðmundur, Zóphanías, Jarþrúður, Ari, and Hrefna's interpretations of glacier change. Perhaps they know just enough to deny the whole.

Lovísa appears at first glance in literal denial. But looking at what literal denial actually entails casts some doubt on this theory. Through a series of case studies on human rights violations, Stanley Cohen argued that literal denial is outright disavowal. Nations or states might claim that they "would never allow something like that to happen [for example, genocide], so it could not have happened," and achieve disavowal by "attacking the reliability, objectivity and credibility of the observer."[129]

Focusing on Lovísa, she does not outright deny that climate change exists, nor that glaciers are changing. She says: "…the glaciers come and go, they are crawling farther farther back and forth. That's the normal thing…"

What Lovísa is doing is normalizing glacier behavior regardless of how out of place current glacier behavior might be. Lovísa also does not deny climate change—rather, she expresses her opinion that it simply is not happening in Iceland: "…the global warming we are not really seeing that here…"

Cohen, again discussing human rights, observed that literal denial is hard to maintain. Rather, states or nations rely on interpretative denial as a viable alternative: "the standard alternative is to admit the raw facts—yes, something did happen: people were killed, injured or detained without trial—but deny the interpretative framework placed on those events… the harm is cognitively reframed and then reallocated to a different, less pejorative class of event."[130]

Lovísa, Jarþrúður, and Hrefna all admit that glaciers are changing, however, they each choose to reject the common scientific narrative—that current glacier change is unprecedented glacier recession triggered by increasing climatic changes. Jarþrúður reasons "that is in a way natural. Because they were growing and growing and growing, so why shouldn't they retreating and retreating and retreating . . ."

The focus on the natural is almost a decoy, a specific "non-focus" on climate change. If glacier change is natural, it cannot be linked to global climatic changes, which are decidedly *not natural.* Hrefna also focuses on the adjective 'natural' when she says: "…some scientists say that's actually what's happening, like recycle of glaciers, they get smaller for some time, and then at some point they are going to grow again…"

Icelandic glaciologists have found and publicized links between climatic changes and recent glacier change. This is not an overstatement; to contextualize, this an island with a small population and a handful of glaciologists.

I talked with the glaciologists working at the University of Iceland and the Icelandic Meteorological Office (MET), the two principal research centers in Iceland. They unequivocally linked climate change and recent glacier change—their issues of difference tend to be more related to rates of recession and future modeling.

The one outlier is the former director of the MET, Páll Bergþórsson, who appears in the Icelandic media warning about Iceland entering a cold period.[131] While regularly labeled and subsequently dismissed as a climate denier, Páll's continuing presence in the media, though not supported by the University of Iceland nor MET, could potentially lead to Hrefna's interpretation that the science of glacier recession is still in question. But this is a generous reading—one voice in a storm of consensus does not lead to an interpretation that all the science is unclear, nor that the change in glaciers that she witnesses *every* day for work, is unsettled.

There is intentional choice here. Like Jarþrúður, the focus on the perceived nonconsensus of science is a specific "non-focus" on the cause of glacier change, a pointed evasion of the primary issue. Evasion means these Icelanders do not have to see, or perceive, glacier change happening before them *right now*—and the consequential baggage attached to present changes such as the likely future disappearance of the island's glaciers—and subsequently are not impelled into any further behavior, emotional response, or practice.

If they do not see it, it might not be happening. And they might not have to do anything.

Norgaard's research also focused on the third type of denial articulated by Cohen, implicatory denial. As Cohen writes, for this type of denial "there is no attempt to deny either the facts or their conventional interpretation. What are denied or minimized are the psychological, political or moral implications that conventionally follow. The facts of children starving to death in Somalia, mass rape of women in Bosnia, a massacre in East Timor, homeless people in our streets are recognized, but are not seen as psychologically disturbing or as carrying a moral imperative to act."[129]

In Norgaard's fieldwork, unless she asked about climate change, it rarely arose in conversation. When it did, it was not denied so much as it was de-localized. What really could the community members in this small town in Norway actually *do* anyway?[128]

In Höfn, climate change was never denied outright; rather, when it was mentioned to me, it was in context of spatially far-off events—the California drought, or Fort McMurray burning, or islands drowning—and critically, it was rarely mentioned in context *with glaciers* unless I as the researcher pushed the connection.

Look carefully at Lovísa's statements. She does not question climate change itself. She just questions whether it is happening in Iceland. Towards glacier change, however, Lovísa, like Friðmundur, Zóphanías, Jarþrúður, Ari, and Hrefna, does question what is specifically happening with the ice. The change is natural, the ice will come back, etc. Such interpretations cannot be explained entirely through denial, nor, as discussed in the previous section, through ignorance.

Denial, in any of the three typologies discussed above, is a useful classification for environmental perceptions and behaviors. Denial as a label helps us to understand the emotional plasticity of interpretation when recognizing glacier change—and, importantly, what those emotions imply, or, more accurately, implicate. It is essential, however, not to stop there. Denial, like ignorance, is part of an open-ended assemblage of relations with glaciers. Perhaps Lovísa, Friðmundur, Zóphanías, Jarþrúður, Ari, and Hrefna are in some denial, but this does not explain everything.

THERE ARE MULTIPLE READINGS, interpretations, and stories of glacier change. Clearly, all people living on the southeastern coast of Iceland do not see glacier change and narrate it in the exact same ways. Glaciers and their movements are contested, argued about, believed and disbelieved.

This unto itself is a critical signal: a large part of the social imaginary relating to climate change and glaciers is underwritten with an assumption of a single story of glacier change—one of loss and melt. As such, it is important to be aware of the social plasticity of glaciers, of their abilities to verify multiple conflicting narratives all at once. Glaciers and how they respond to changing environments are proof for all sorts of different stories—dependent in large parts on who is telling the story.

This is, of course, not limited to Iceland. For one example, anthropologist Georgina Drew conducted ethnographic research in Garwhal, India, examining hydroelectric dams. In the course of her research, however, she also paid attention to how local people perceived the Gangotri–Gaumukh glacier and its runoff feeding into the dam reservoir.[22]

Drew found that some people in the area comprehended glacier change not as a symptom of global climate change, but as part of a larger cosmological action, and, critically, were not perturbed that the glacier or the river it fed might disappear because, for them, the goddess Ganga would always be there—in the glacier and in the river.[22, 23, 132, 133] Said another way, Drew found that local spiritual frameworks were often incompatible with the dominant scientific framings of Himalayan glacier recession. She showed how indigenous knowledge of the surrounding environment was informed by specific Hindu texts and cosmologies that taught certain destinies regardless of human action; in this case, that the goddess Ganga (stationed within the glacier and river) might indeed disappear from sight, but "she will continue to exist in the heavens and in the underworld."[22]

As such, Drew did not find the people she interviewed ignorant, uninformed, or in denial, rather, she argued that their glacier change narratives were influenced by local cultural practices and norms.

In Höfn, a cultural refrain I heard over and over centered on the sheer survivability of the Icelandic people, and how this characteristic was especially relevant in Hornafjörður, where people over the centuries had to survive living with glaciers. To survive was regarded as part of Icelandic heritage, a part of Kirsten Hastrup's "Icelandicness."[52]

In Höfn, modern life is still relatively new, but it is a place where individual and communal memories stretch back just a few decades to harder times, to turf homes, food insecurity and scarce medical interventions, and threats of glacier surges and volcanic explosions. Looking at the community today—the orderly streets and modern cars, the solid-built homes and visible infrastructure and social services—it might be easy to forget just how recently this society lived on the brink, where acts of nature could drastically redirect a person's and a community's survivability.

This cultural memory of previous times informs today both how people interact with each other (a general aversion to conflict) and how they interpret the local environment in part through the lens of survival. This is essential to understand: glacier change can be conceived of as a threat to survivability—not in isolation (the change of an individual glacier), but because of the baggage glacier change (collectively) may bring. For Icelanders, survival can be an intergenerational characteristic in the face of continual precarity.[134, 135]

Here, in this place, interpreting survival involves interpreting glaciers—it always has. As Leifur from the previous chapter, the retired man in his late 90s who spent two months digging a new channel for the Hólmsá in 1937, said: "We always thought about glaciers."

Leifur meant that for the people living in the region, glaciers permeated their daily lived experiences, their everyday lives, and they had to each day consider how to survive the ice. The power glaciers exerted determined fundamental survivability for people in this region, and Icelanders were *always* aware of the glacier's power.

Leifur's comment was supported by many other local Icelanders, who regularly discussed concepts of survival with me during interviews. For example, Þórður, an local man in his 40s, observed, "The atmosphere here in this society is to survive. I believe that because of the generation before, and the generation before that, they had a very strong survivor thinking. So I believe it exists here still."

Another local man, Diðrik, built on Þórður's comments, observing that he "could understand why it is interesting for people to go to the glaciers because of this magnified feeling when you stand near it and on it . . . but, the people from earlier times in Iceland, they had to survive. They had no time for fun. So they just had to survive and they took the sensible way of meeting nature, of not meeting it if it was possible to not do so."

Dadda, a woman in her 30s who recently moved to the area, explained to me the essence of Icelandic survivability: "I think this is part of the Icelandic soul. The root is that we always have bad weathers in Iceland, in the last decades, for hundreds and hundreds of years. We have had to adapt to all kinds of weather, and the harsh nature and not having so much to eat, and we always had to find a way to figure things out."

Many other Icelanders of all professions spoke about concepts of survival. I only noticed local Icelanders under the age of 25 rarely mentioning anything regarding the hardship those who came before them experienced.

Oral history tells stories of how the glaciers always oscillated back and forth. When the ice shrank, Icelanders counted on glaciers growing bigger; when the ice grew, the people knew it would eventually shrink back. This was the expected rhythm of glaciers, the normalization of a volatile process. Overlay this narrative—expectation—of glacier change with Icelandic cultural legacy: generation after generation of Icelanders, able to survive in tandem with the ice. The ice shrank, people on the southeastern coast adapted and survived. The ice advanced, people on the southeastern coast adapted and survived.

Here, there is an *expectation* of over a thousand years of survival alongside the ice.

Such expectation emerged clearly in the way people spoke about living in Iceland today. For example, Árvök, an older woman who was born in Höfn, described living with oscillating glaciers. She told me, "Here you have the glaciers. It is just a part of you. Like the nature. This is just what you are used to, and it has always been there, and you are worried and concerned about these changes, but there is not so much that we can do. Even though we want to. It is just like, I think, like living near volcanoes. It is just a part of the life, we just have to adjust and learn how to live, live there with the glaciers, and know when to leave. We are concerned, but it is Iceland. What can we do? We just learn to live with what we have, and that is what we always do. There are constant changes here and you have to live with them. It is just the way it is. I think Icelanders, they are so used to the changes, that they just try to adjust. That's the only thing you can do or you move away from Iceland."

What Árvök clarifies is explicitly brutal: in this place, a person either adapts and survives glaciers, or they leave Iceland. There is an expectation in Árvök's view that this is the way it is, the way it has always been—people survive the landscape.

Þórður, the man who previously observed that it is the nature of this society to survive, expanded his point when he spoke about how with isostatic rebound (land rise associated with receding ice) and increased sedimentation from the glacier rivers, Höfn's harbor is losing its access.

Höfn has the only viable harbor for hundreds of kilometers on the south coast. But, as the glaciers melt, the land is springing back up, causing the depth of the narrow passageway into the village's harbor to become shallower each year. Boats weighed down with freshly caught fish drag their bottoms on the sandy sea bed.

"In waves one to three meters high," Þórður explained, "the pelagic boats cannot enter the harbor. If the boats are full of fish, we can wait for twenty-four hours, and often we do that. We try to manage our fleet so they are not stuck here because of the waves outside the harbor, or keep them outside the harbor and ask them to catch when we see it getting realistic to get into the harbor again. This is about administration and control of the fleet and it is workable. But this, talking about weather and landscape and nature, the life in Iceland has always been part of, you know, to survive. To have a good life this is all some kind of adaptation to the nature. For the fisherman, for the farmers, for everybody."

Þórður said that dealing with shallower and shallower harbor entrance is just like all things in Iceland: people have to survive. They adapt and survive, or they don't. Right now, the harbor is getting shallower, and this is pushing the fishing companies to adapt. They have dredged the harbor, they will dredge again. The company is using technology to maximize wave prediction to help boats enter the harbor when the water levels are the highest.

There isn't any choice—and this has been always part of the life in Iceland. While the island has always changed, people have survived *with* the changes. People here have transformed with transforming landscapes.

Survival is normalized, much as regular glacier change is. I did not find evidence in any of the interviews I held with locals Icelanders—a fraction of which are quoted in this chapter—to support claims they were ignorant, uninformed, or in denial about glacier changes.

Yet narratives contrary to the material processes of glacier change multiplied, and I heard them from Icelanders all over the region.

What I think is happening here is this: Lovísa, Friðmundur, Zóphanías, Jarþrúður, Hrefna, Ari, and other Icelanders are influenced in part by past expectations of survival. As a way to make sense of today and an unknown future, they negotiate it through controlled expectations of what happened in the past.

Look again at Lovísa's comments: she made the point that "the glaciers come and go," and that is the "normal thing." Icelanders, she said, are used to change.

Implicit here is an expectation of change. What is happening today in her view is nothing extraordinary and so she makes sense of local glacier change by augmenting her incomplete knowledge with cultural knowledge, a past experiences of survival and change. "It is very hard for somebody to convince an Icelander that global warming is real, because we are really not seeing it," Lovísa said.

Lovísa is not denying change—but rather pointed to the pattern of change that *always* happens in Iceland.

Jarþrúður drew on Helgi Björnsson's lecture, focusing on his discussion of widespread glacial recession in the latter part of the 1800s. She found relief in not only understanding that she was not at fault that the ice was melting, but also that it was a natural process that had happened before. Implicit here is that *it happened before.*

Jarþrúður's family survived glacier recession in the 1800s, so too will she—and all her grandchildren—survive it now. Jarþrúður normalizes what is occurring by narrating glacier change as a natural, re-occurring process. This positions today's events into as survivable and safe.

Even Ari does this. He believes that in the next fifty years, *something* will happen to reverse glacier recession. He believes this in part because, looking at Icelandic history, this has been the pattern of glaciers—oscillating back and forth, back and forth. Ari is narrating the future through past experiences with glaciers.

But the extraordinary nature of modern glacier change, of Heinabergsjökull's unusual recession, of glaciers disappearing with no evidence of speedy returns: some Icelanders in this region articulate views of glacier change contrary to these documented material behaviors of ice. I think they do this as a form of survival. This is, consciously or unconsciously, a process of self-protection predicated upon a cultural fabric woven with the thread of survival.

For some, this centers on unseeing, on negotiating glaciers into easier spaces of psychological safety that includes not necessarily an untrue narrative of ice so much as an incomplete one. Or, some people might re-focus glacier narratives onto the behavior of one glacier such as Heinabergsjökull, or on supposed scientific controversies, or on ice returning at some near point in the future—all to protect themselves, survive, and to move forward into safe futures.

A LAST POINT TO CONSIDER, but important.

I also suspect one additional variable is occurring that influences how people perceive glacier change in this area.

Glacier change—and climate change more broadly—is regularly narrated (talked about) in a reductionist way that reduces all the potential futures out there down to a single future—one that is certainly negative and ice-free. You've heard those stories: we are losing our ice and by such and such year, a particular place on this planet will no longer have glaciers. I've written many of those stories over the years.

Here's the problem. An ice-free Iceland of the future would be a hard place to survive for Icelanders. Any place that has ice today would be a hard place to be without ice. Glaciers are critical to local ecosystems. It is terrifying to think about a future that has no ice. And as such, the sheer "unsurvivability" of an ice-free Iceland might to some render that future unthinkable.

We simply choose not to think about it.

A local artist, Rúnar, phrased it best when he said: "My bronze sculpture will be there longer than the glacier . . . sadly I think that they are going to disappear, and it is also unthinkable."

I think that the unthinkability of loss plays a strong part in answering why some Icelanders produce, and cling to, contrary narratives of glacier change. It is easier to make a different story of what is happening than to take a hard look at what is quite unthinkable.

Unthinkability as a social phenomenon has not been treated much within climate change scholarship, but it has been a focus within studies of war, genocide, and international relations. But to talk about glaciers, we also need to talk about war. Bear with me one moment—this is important.

Timur Kuran, an economist from Duke University, explained that unthinkable thoughts are thoughts one cannot admit having or entertaining "without raising doubts about one's civility, morality, loyalty, practicality, or sanity."[136] To this list I suggest adding survivability. How may we speak of our own personal inability to survive without questioning our sanity? For example, those who entertain notions of suicide are regularly directed to mental health services. Self-harm for some is unthinkable—and thinking thoughts of self-harm are strong signs that something is amiss. Mental health professionals do not want people to have such thoughts.

Or, to use a different example: it is unthinkable that I, or you, would willingly commit genocide. Kuran, again: "An unthought belief is an idea that is not even entertained."[136] In this theorization, unthinkability is an emotional concept. To Ari, glacier change is unthinkable because a world without ice is too horrible to consider—therefore, he can skip over the present and discuss his certainty that the ice will return.

Unthinkability is not necessarily synonymous with improbability; for example, as political scientist Jennifer Mitzen observes, when countries label nuclear war "unthinkable," the concept is deployed to suggest that while the likelihood of unleashing nuclear weapons *is* probable, it is too horrible to consider.[137] Unthinkable thoughts are also those that are ontologically not possible.

For example, it is unthinkable that a person walks through a wall. It is imaginable, but a person cannot do it—so unthinkable pertains to things that are real to people within personal realities.[138] For most realities, walking through walls does not register.

Crucially, realities can be mistaken. Friðmundur and Zóphanías are mistaken about Heinabergsjökull, but this is within their reality. Even if pressed for evidence of glacier growth, what matters more is their reality: these two men believe the glacier is growing and as such, they interpret local narratives accordingly.

Unthinkability may also seem like denial, and people can deploy the concept as a strategy to avoid confrontation. When events are labeled unthinkable, people may be freed of the responsibility of speaking about them or assuming responsibility for the acts themselves.[139] This aligns with Felix Frankfurter, the Jewish Supreme Court judge who heard eyewitness testimony about the Belzec concentration camp, and responded: "I am unable to believe him. … I did not say this man is lying. I said I am unable to believe him. There is a difference."[112] The genocide that happened day after day after day in the concentration camps was unthinkable.

Jarþrúður was relieved to understand that glacier recession was natural—she overlooked the majority of what Helgi presented that day and instead clung to a reality in which she does not bear responsibility for glacier loss. For some Icelanders, glacier change—the loss of the southeastern coast's outlet glaciers—remains simply unthinkable. As such, other things come into focus.

People focus on the specific details of one glacier out of a group, or on perceived scientific contestations, or when the ice might return, or on disavowal of present change. Lovísa, Jarþrúður, and Hrefna focused on the mechanics of change—natural or not natural. It is easier to focus on those mechanisms than on a potential outcome. Ari focused on the return of the ice, overlooking evidence that suggests glaciers may be too far gone. Friðmundur and Zóphanías repeated narratives of Heinabergsjökull growing.

What unites these disparate interpretations is the undergirding of survivability—the ability to move safely through each day, to move forward into an imaginable, dependable, and safe future. This is about survival, about normalizing an uncertain landscape, about reaching for past expectations of survival and conditions to make the future more controllable.

Unthinkability is predicated upon Icelandic cultural notions of survivability. What is happening in Hornafjörður is a process of self-protection predicated on the force of cultural narratives of environmental change. People will, consciously or unconsciously, shield themselves; and in Iceland, part of the cultural fabric is woven with the thread of survival, which for some is centered on *unseeing*, on negotiating glaciers into easier spaces of psychological safety that include not necessarily an untrue narrative of ice so much as an incomplete one.

George Marshall wrote that he believed climate change was the "ultimate challenge to our ability to make sense of the world around us. More than any other issue it exposes the deepest workings of our minds, and shows our extraordinary and innate talent for seeing only what we want to see and disregarding what we prefer not to know."[112]

Stories—and interpretations of events and phenomenon—change over time. Likely many of the people featured in this chapter have developed, expanded, or evolved their own interpretations of glacier change from the time of my fieldwork. And it is this that is one of the points at the heart of this book and this chapter in particular: there is no single story of anything, and stories are not static. The stories we create are profoundly shaped first by who we are as a people over what we see before us. Metrics never tell the whole story.

Today, glaciers very much still exist. But it is difficult to "not see" glacier recession. Glaciers carve landscapes in their image; when they disappear, they leave empty casings that shout hollow stories of the ice that was once there. And yet to some people, glacier recession remains invisible, unthinkable—not because of ignorance, or lack of information, or denial, but rather because it threatens individual and community senses of survivability in Hornafjörður.

The thing is, at some point in the not-too-distant future, the entire trough-bound tongue of Heinabergsjökull will collapse.

Instead of the glacier receding incrementally backward in concert with the region's other glaciers, the whole floating tongue of the outlet glacier will crumble apart, leaving behind a fourteen-mile-long vale full of water and chunks of glaciers.

It will be as if someone cut the glacier off at its base, as if someone cut a person's tongue off at the base of their mouth; left the tongue floating silent to decay down into pieces and eventually dissolve away.

CHAPTER SIX

≈ ≈

"Glaciers are coming to life, just now, in their twilight hours."

J.B. MACKINNON, 2016[140]

"I just felt like this was, I felt this old, wise energy. That was what came into my mind, this was so old and this is so wise. It knows, this phenomena knows the stories of thousands of years."

SOFFÍA, 2016

LATE AFTERNOON AT THE END OF NOVEMBER, I drove forty-five minutes west from Höfn through winds determined to shove my small blue car off the road. At a nondescript pullout lined by rocks, the intersection of the Ring Road and Road F985, I parked my tiny Volkswagen Polo, hopped out, and hopped into the waiting Icelandic super jeep.

My friend Jökull was hunched over the driver's seat cradling a cup of hot coffee. Jökull lived in northern Iceland, but he was down on the south coast looking for new ice caves. He'd been waiting the last forty minutes for me to meet him.

Jökull handed me the coffee, put his truck in gear, and we drove slowly up the F-Road. In Iceland, there is a special classification for certain mountain tracks called F-Roads. These roads are typically unpaved, poorly graded, un-bridged, all with varying levels of accessibility. Generally, F-Roads become accessible in the middle of summer and require 4-wheel drive vehicles.

The F-Road we were driving on was a road in suggestion only. It was a rutted and exceptionally steep track with sharp drop-offs on the opposite side from where the road was cut out of the mountain. As we wound higher and higher, away from the ocean and the flatlands of Mýrar and Nes, I kept my eyes directly on the road in front of us. Looking over the side made me queasy.

Jökull laughed at my green face. He was in high spirits. This road was just another Tuesday to him, and that day, we were headed to Jökull's favorite glacier.

Jökull and I have known each other for over six years—he helped me with some expeditionary work in northern Iceland several years back and we remained friends, exchanging annual Christmas cards, social media updates, and ridiculous pictures of cats and glaciers. Jökull was in his mid-thirties, slender, wore a plaid cap permanently affixed upon his head, spoke English with a curiously thick Scottish accent, and was quite involved in the island's burgeoning environmental movement. He and his wife owned a tourist business outside Akureyri.

When I told him about my research on the south coast, he promised to personally introduce me to his favorite glacier on the island. After months of juggling schedules, we had finally found a time: now.

Upwards we drove, and drove, and drove, snaking alongside steep escarpments of volcanic strata cleaved by ice, water, and machines.

Much sooner than I predicted, Jökull parked the vehicle in the middle of the road—a classic Icelandic parking job—left it running, and gestured for us to hop out. The winds that had threatened to blow my little car off the road were only worse up at higher elevation, and the light was weak and watery.

We left the jeep and walked in strong winds for fifteen minutes, navigating over bare rock and periodic lichen and moss colonies. I could barely see as the wind kept blasting icy and sandy wind into my face. But then, like magic, we trudged around a jagged rock outcropping that blocked the wind, dropped three feet down onto a comma-shaped ledge, and below us stretched the enormous tongue of Skálafellsjökull.

Skálafellsjökull is a sleek and languid glacier. Start to finish, the glacier sweeps over fifteen grey-blue miles beginning at the Breiðabunga plateau and ending in a small proglacial lake at Hjallar. Aerial images show a broad and glossy accumulation zone that gradually narrows as it funnels east off the Vatnajökull ice cap.

The flow of Skálafellsjökull is mostly southeastern, and, as one woman observed to me, as even as a cat's just-licked fur. Only towards the terminus of the glacier do things go astray. There the smooth ice collides straight into the headwall of the mountain Hafrafellsháls, which forces it to switch directions and pour south around the mountain, pivot east, then stream full steam straight into the rigid east-facing valley alongside the mountain Skálafellshnúta. The steep escarpments on either side constrict the body of the glacier, compressing it like toothpaste through a two-mile long narrow valley where it finally terminates in ripples of pressurized crevasses.

From our wind-protected perch, we could see most of the glacier's final two miles.

Jökull told me he came there, to Skálafellsjökull, first with his father years ago when he was a kid. His dad worked a short job over several weeks nearby, and while his dad worked, Jökull ranged freely during the long summer days. He eventually explored the glacier.

And, he told me, one day when he got too close to the ice, he heard Skálafellsjökull snarl. "Like a dog before a fight," he described. All the hairs on his arms stood up. "I knew [then] that Skálafellsjökull was very powerful. That the glacier had breath coming in and out, and that the glacier knows I am here."

He spent more days with the glacier, and he said he gradually came to know that it was watching him, that it was aware and conscious of him walking nearby. He also started to sense that Skálafellsjökull radiated a certain power.

When young Jökull questioned his father about his experiences with Skálafellsjökull, his dad told him the glacier was alive.

I asked him to explain.

Jökull told me Skálafellsjökull was his favorite glacier because, while he had grown up hearing Icelandic stories about sentient rocks, rivers, glaciers and much more, Skálafellsjökull was the first time he experienced, on his own, a *living* glacier.

Today, Jökull regarded the glacier as a friend he visited when he was in the area. From his perspective, Skálafellsjökull was an individual creature, living and conscious.

For over an hour, Jökull and I sat there in the small rock hollow tucked into the mountainside, shoulder to shoulder, looking down on the miles of grey-blue frayed ice as the wind wove through the glacier's crevassed hide, and he told me stories of all the times he'd visited Skálafellsjökull. At times he fell silent, and we just sat there and watched Skálafellsjökull and listened to the wind, the rocks, the glacier.

Before we packed up to leave, Jökull casually told me he thought the glacier was "eating itself from the inside out."

It took me a few minutes to understand what he was telling me, to contextualize what he described alongside what I had heard from many other scientists.

Skálafellsjökull since 1890 had retreated 1.2 miles,[40, 41] and the glacier is in what is informally referred to as a destructive feedback loop. In essence, the glacier *is* eating itself.

What is happening boils downs to this: the act of Skálafellsjökull's recession causes the rate of Skálafellsjökull's recession to increase. As temperatures increase and glacial ice melts, meltwater accumulates on the surface, sides, and under the ice—which then transforms a glacier's albedo (how much of the sun's energy is absorbed or reflected out). When the frozen glacier body is covered with glacier meltwater, it increases (exponentially) how quickly the ice melts.

This process befalls many other glaciers in Iceland, and glaciers worldwide, what geologist Marco Tedesco from Columbia's Lamont-Doherty Earth Observatory phlegmatically described as "melting cannibalism, basically—it's melting that's feeding itself. . . . Rising temperatures are promoting more melting, and that melting is reducing albedo, which in turn is increasing melting."[141] Hence, like many glaciers in the world, the more Skálafellsjökull melts, the quicker Skálafellsjökull melts.

Outside of Iceland, two other notable examples of glacier cannibalism include Alaska's Muir Glacier, which through escalating recession between 1893 and 2004 has lost over thirty-one miles of ice,[142] and the remaining glaciers atop Kenya's Mt. Kilimanjaro, which are estimated to be entirely gone by the year 2040.[143, 144]

To Jökull, it is almost unbelievable how much of Skálafellsjökull has materially disappeared over the length of their friendship.

He pulled out his phone and showed me images of Skálafellsjökull from five, six, and ten years previous. The change in the glacier was easily discernible. Jökull said that he worried about his friend Skálafellsjökull—that each time he visited Skálafellsjökull, the glacier looked decrepit and sick. It reminded him of how his dad looked when he went through chemotherapy.

We drank coffee quietly together as Jökull explained to me how his father died, and why he connected that to Skálafellsjökull. What struck me as we talked was how clearly Jökull perceived a livingness, an inner personality, a force of life, emotions, and consciousness in Skálafellsjökull—he sensed something intangible in the ice that was as real to him as his father.

Jökull was not alone in his perceptions of living glaciers. Many Icelanders I talked with in Hornafjörður from across social groups—men and women, old, middle-aged, and young, sailors, scientists, and seamstresses—spoke about glaciers as in some way alive, in some way sentient. What that aliveness looked like, how that sentience was percieved, was different for each person and expressed in mixed, and at times, antithetical, ways.

While not all Icelanders I interviewed conveyed perceptions of glaciers as alive, many did, and they would talk about glaciers in much the same way they spoke about other people. Icelanders talked of visiting the glacier, loving the glacier, fearing the glacier, worrying about the glacier, mourning the glacier. People told me the glacier was watching, feeling, interacting, blooming, breathing. That the glacier possessed cognitive awareness; that the glacier did not have enough to eat. They said glaciers had individual needs, had pasts, held memories, fought, had will, physical movement, ability to transform and subsume, awareness, temporal characteristics. Locals frequently described how glaciers were growing or dying, arriving or departing, sleeping, resting, healthy, not healthy, chattering. Some people told me they thought the glaciers were happy.

This aligns with a faint current I've noticed in some circles lately, an emergent sense of glaciers reviving—waking—both in terms of public consciousness and academic scholarship. As Canadian journalist J.B. MacKinnon observed, "[s]trangely (or again, maybe not, depending on your perspective), glaciers are coming to life, just now, in their twilight hours. For years, the predominant view has been that they are not only lifeless, but hostile to life."[140]

The predominant view, I would hasten to remind, has also been focused on a specific glacier process: melt. Glaciers are widely viewed as things melting: lifeless and hostile to life. So much so that the notion that they might be perceived as not only alive (the antithesis of lifeless) but also involved in life (your life, my life, a bird's life, all the lives of those living within the community of Hornafjörður)—goes directly against the grain, against prevailing views, against what many people might presume is true of ice.

But there are many diverse interpretations, perspectives, experiences, and knowledges of glaciers across the world today, including Jökull's and many other Icelanders, and I suspect it might be rather revealing to consider what perspectives of glaciers alive tells us about how people and ice interact today. More broadly, thinking about glaciers alive invites us to radically re-think, re-imagine, and re-possible both what glaciers are, and what life *could be*. It also dramatically re-orients how we might think about environmental change and what is actually at play when we talk about glacier loss across our planet today.

As journalist Robert Kunzig writes, "glaciers are wild beasts . . . they breathe . . . they move . . . they rule[d] . . . they struggle,"[145]

. N .
V ✦ A
· S ·

IMAGINE, for one moment, a glacier *alive*.

What are you envisioning?

Let me right on the outset disabuse any vision you might have that a glacier that is perceived alive or some variant of living by some people is a glacier with comically sad eyes and ears standing on two melting legs looking at you. This is not what people sense when they sense sentience—aliveness—in glaciers, or really anything else under our sun.

Perceptions of glacier sentience vary worldwide. Julie Cruikshank recorded the life stories of Indigenous women in Canada's Yukon Territory, British Columbia, and southeast Alaska living near glaciers and found that in local perceptions, glaciers were not sentient because humans willed them so (think of anthropomorphism); instead glaciers were autonomous beings with emotions, awareness, and varying interest in human affairs.[18]

In northern Italy, some German-speaking villagers understood glaciers as sentient. Villagers spoke of the "beginning of the dying" as glaciers grew "naked," and they felt the ice should be "left in peace."[146]

In the United States, people living in the shadow of Mt. Shasta, a distinctive volcano in northern California, attributed an *awareness* to the mountain and its glaciers and observed the mountain's ability to protect itself by triggering icy avalanches.[7]

Perceptions of glacier sentience have also been reported at different scales and degrees from diverse places such as the Karakoram,[147] India,[22] New Zealand,[148] Bolivia,[124] and the Greenland ice sheet.[149]

Broadly defined, sensing an aliveness—what I am going to term sentience for the rest of this chapter for ease of explanation—is to sense the ability of *any* thing to think, feel, or be aware, independent of humans or other things. Thus, when sentience is perceived in something, be it a glacier, a tree, or a cat, it is that thing's livingness that is perceived. It is not that thing's appearance of sentience (the sad eyes or melting legs), it is the livingness, the aliveness, within that thing.

It is not useful to focus on trying to prove sentience, nor to develop a metric for levels of sentience perception. People do not determine how they're going to move through our world based on strictly defined classifications of what they perceive is alive and what they perceive is not. As anthropologist Tim Ingold contends, people worldwide rarely agree about what is alive and what is not—and typically do not limit themselves to thinking and experiencing the world through scientific predetermined biological and physical classifications.[150]

That being said, it is helpful to first work through some of the vocabulary and terms associated with sentience. Several concepts entangle with sentience, including anthropomorphism, personification, teleology, and animism that can overlap in confusing ways. Let me explain each briefly so we can have a level playing field moving forward into sentient glaciers.

Anthropomorphism is where a thing is doing something human-like. For instance, when a glacier hugs another glacier. Personification is where a thing *appears* like it is doing something human-like. Like when a glacier with a crescent-shaped terminus appears to be smiling. Anthropomorphism and personification typically revolve around detection of various degrees of human-like features or characteristics on elements that make up the world around us, such as frowning animals or suspicious house plants.

Teleology, a less widely used concept, ascribes purpose and rational decision-making onto a thing. For example, a glacier might choose to recede. Or, a cat might choose to knock a plant off a windowsill because you did not provide adequate breakfast.

Last, animism. This is where a thing is perceived as living and conscious independent of human beings. For example, a glacier with its own point of view. Anthropologist Tim Ingold noted a commonly held and rather shallow understanding of animism as those who project animation (read: life) into passive or perceptually lifeless objects—for example, a human animates a glacier.[150] In this interpretation of animism, any person—you, me, your cat—could *will* life into a glacier and subsequently bring it to life.

This is, however, a human-centric application of the concept. Ingold argues this is not how animism works. Animacy, he writes, "is not a property of persons imaginatively projected onto the things… it is the dynamic, transformative potential of the entire field of relations within which beings of all kinds, more or less person-like or thing-like, continually and reciprocally bring one another into existence."[150]

In animistic perspectives, things are alive with or without people. Things are not defined relative to human beings; they do not require a person to perceive them for them to be alive.

I linger here so you linger. Animism is a complex way of being in the world, a sophisticated ontological position—how a person perceives their fundamental nature within the wider world—which is predicated on diverse relations with the environment and the world as a whole. As religious studies scholar Graham Harvey observed: "[a]nimists are people who recognize that the world is full of persons, only some of whom are human, and that life is always lived in relationship with others."[152]

Perceptions of aliveness in the surrounding world are heavily associated with Indigenous, First Nations, and Aboriginal Peoples. For example, geographer Mabel Gergan's thoughtful article on post-human geographies, she explains: "I use the term "sentience" or "sentient" to specifically reference the indigenous belief in the quality of all life to think, feel, and act."[153]

Gergan positions "indigenous" as a defining characteristic *of* sentience.[153] For something to be perceived as alive in her definition, it has to be perceived as such by an Indigenous person. Non-Indigenous perceptions of glacier sentience, such as those from Iceland, are outside of Gergan's definition (so too might be perceptions of glacier sentience reported from Italy,[146] New Zealand,[148] or northern California[7]).

Substantial research reports that many peoples from Indigenous, First Nations, or Aboriginal Peoples' communities do have animistic understandings of their environments, including from Australia,[154, 155] sub-Saharan Africa,[156] Amazonia,[157, 158] and especially the circumpolar North, where the apparent livingness of things fits within the worldviews of many Arctic and sub-Arctic peoples.[150, 159] In the North, "natural elements such as glaciers, mountains, seas, and animals are seen as sentient beings, having agency, emotions, and interest in human affairs."[160]

Importantly, however, such research does not signify that perspectives of sentience are limited to just these communities. Rather, it is much more likely that research focused on animism and sentiency within these specific communities because it has been widely assumed that people in Western societies do not perceive sentience outside of predetermined biological and physical classifications.[150, 161] As in, some people might not examine animistic perspectives within my community in Oregon because this community is a "Western society." Said another way—you don't see what you are not looking for, and often researchers are not looking for perceptions of sentience outside of Indigenous, First Nations, and Aboriginal Peoples' communities. Subsequently, perspectives of sentience and/or animism are less likely to be explored within in places such as Iceland.

This is important to think about because, contrary to common assumptions, perceiving something as alive that is more than human is not just within the prerogative of Indigenous, First Nations, or Aboriginal Peoples. It could be perspectives held by you, me, or anyone worldwide.

Interestingly, in places where sentience is perceived in more than human things, sentience is usually conceptualized *beyond* specific environmental phenomenon. For example, peoples of Arctic and sub-Arctic regions do not limit sentience recognition solely to glaciers; in accordance with their worldviews, recognition extends to the surrounding environment and its contents, including other peoples and other knowledge systems.[18, 162-164]

So, when a person perceives a glacier as alive, likely they're not going to believe that just the glacier is alive. Likley they'll also extend those views to include, for example, trees, rivers, or clouds. Many Australian Aboriginal Peoples include *entire* landscapes as "a sentient partner[s] in the experience of being human."[154] So when environments are understood as living or sentient, to some people that implicates those environments and landscapes as fellow partners on this planet.

We are now starting to get close to seeing why perspectives of sentience are quite important today to consider. It is not the perception alone—the understanding of a living Skálafellsjökull—it is also what that perception implies, entails. It is what comes *with* that perspective.

Part of perceiving sentience might also include perceiving a partnership—perhaps even a responsibility for and of that sentient thing and that sentient landscape that that thing lives within. It might include caring about that glacier the way you might care about a family member or a friend. I might include caring about that environment that you and the glacier live in together, as partners.

PEOPLE DID NOT COME RIGHT OUT and just tell me they thought various glaciers around Hornafjörður were alive. Research rarely happens in such a cut and dried manner. Rather, during sit-down interviews across hundereds of homes in the area, in focus groups and ride-alongs, while shopping at the town's grocery store, during presentations and late evenings over coffee with friends, sentient glaciers continuously murmured in the background current.

Like the previous chapter, where some Icelanders believed that glacier recession was not happening, and some people believed that it was, I found equal diversity in how some Icelanders perceived sentience in glaciers. Some Icelanders thought glaciers were alive, and some did not. Of those that did, how glaciers revealed their aliveness, or how people perceived sentience, was quite diverse.

Some young people spoke about glaciers as alive, others did not. Both men and women discussed glaciers as living and breathing. Many Icelandic scientists spoke with a fascinating duality about glaciers as clinical objects of study and as friends with whom they maintained relations over the years. Artists, shopkeepers, fishermen, bureaucrats, farmers, tour operators—over and over, people from diverse professions talked to me about glaciers as imbued with livingness, as sentient.

If I asked direct questions *about* sentient glaciers, often Icelanders would either shake their heads in disagreement, divert the conversation, or speak of a glacier's physical or aesthetic properties. But sometimes, if I asked after a specific glacier, or encouraged the talk to keep going, a glacier's livingness might seep into the conversation in different ways. Some people talked about glaciers as partners similiar to the ways Australian Aboriginal Peoples speak about entire landscapes more generally. Other people described glaciers as protectors or watchers, and still others spoke with concern about the vulnerablity of glaciers' life cycles and what they could do to help.

. N .
V — A
· S ·

IN EARLY WINTER, I met up with Soffía, an Icelandic woman who had recently completed her doctorate at the University of Iceland. In her thirties, she grew up in western Iceland on the Reykjanes peninsula with views of the 700,000-year-old stratovolcano and the glacier Snæfellsjökull at its top.

We met at a coffee shop near the University of Iceland, and Soffía was brimming with excitement at having just finished her doctorate degree. She had no job or income, but she was visibly relieved to be done with school.

As we talked, the conversation moved in and around glaciers. Soffía told me general scientific facts about the ice, similar to the ones many Icelanders say, including, for example, how the island's icefield is the largest in Europe.

We carried on for twenty minutes or so about the academic aspects of ice. Then the conversation switched in tone, and she told me she was concerned about how the glaciers were changing on the island, and how people did not seem to be aware. "No one listens to the glacier talking," she lamented.

I noted her language shift from academic to personal, and I asked her to tell me a favorite memory of a glacier.

Soffía smiled and told me a story about Snæfellsjökull, a glacier on the Snæfellsnes peninsula in the west. "I have watched the Snæfellnes glacier since I was ten!" she explained. "I have always been intrigued by the glacier, and there is some kind of energy that comes from it. I grew up with that glacier. But I never touched it. I first went on Snæfellsjökull when I was twenty-seven, but I never had touched the glacier before. But I had looked at it every day since I was a small child."

She continued to tell me how she could always see this glacier throughout her childhood. It was her companion. "When I played outside it was there over the water with me."

Soffía then told me a glacier story about a different glacier—Hoffellsjökull. Several years ago, she went on a trip out to that glacier, and she had an unforgettable experience.

"We went a little bit out on the glacier. You know, it was before there was a lagoon, there wasn't a lagoon at that the time, in 2008 or 2009. And we came up to the front of Hoffellsjökull, and we could just walk out on it," she said. They did not have to go around to the side of the glacier like people do today because of the large glacier lake in front of the ice.

Soffía described, "I just remember this strange feeling when I was looking down and getting on the glacier, and then I looked up, and there was just this huge chunk of ice just towering over me. And my heart just ahhhhhh, just stopped, and I just felt this old, wise energy. That was what came into my mind, like this glacier was so old and so wise. It knows, this phenomena knows the stories of thousands of years. It was this very strong feeling that grabbed me there."

Soffía continued and discussed her experience of being aware of Hoffellsjökull alive, much like she sensed an energy from Snæfellsjökull when she was growing up. Shrewedly, she pointed out that the awareness of their aliveness was not required by her—that it was "beyond me, actually." Rather, Soffía was just struck by their aliveness.

IN LATE SEPTEMBER, I visited Ragnar, a lifelong resident of Höfn, at his home filled with children and animals and countless cups of coffee.

I walked into his home and was immediately surrounded by four dogs, two small children, and a thick cloud of cigarette smoke. Ragnar was excited to see me, and he led me to the kitchen speaking a mile a minute. As he poured coffee, he recounted to me various stories about different glaciers in the area. The story about a man who fell into a crevasse and sang songs until his rescuers found him. The story of a man who lost all his sheep when they crossed the glacier, only to find them perfectly safe a week later grazing in a high mountain pasture. The story of seeing two Arctic foxes chase each other across Skálafellsjökull. The story of tourists crossing the glacier, refusing search and rescue help, only to need it several hours later when a storm blew in. The story of a flood arriving so suddenly it skimmed across the winterized land. Ragnar told me story after story about glaciers—where the ice was the stage upon which all the action occurred.

Only at the end of the afternoon, as I prepared to leave, did the conversation shift.

"I tell people that the glacier is alive." Ragnar told me,

somewhat as an afterthought. He'd exhausted all his other glacier stories. "You can see it, you can feel it, but when you step up on the glacier, it is moving. People find that really strange."

"Why do you think the glacier is alive?" I asked him.

"I have seen it so many times. That's why I say to people, never trust the ice. Not completely."

"But you said it was alive?" I followed up.

"Yes. I have felt it, you know, and I don't know the words but I know the glacier is alive. I have felt it, you know. Felt it."

The conversation moved on. But, standing on his doorstep at the end of the afternoon, I tried again and asked him what a living glacier felt like. He didn't answer.

We parted ways, and I saw Ragnar periodically over the next several weeks but not long enough to continue our conversation. Several months later, I ran into him at the turn-off to Skálafellsjökull. He had suffered an immense personal loss over the winter, and I gave him a hug. I told him how sorry I was. As we hugged, Ragnar leaned down and whispered in my ear, "The glacier is watching her."

He walked away, and I was left puzzled.

A couple of weeks later, still confused, I sought clarification with a friend of Ragnar's named Fríða.

Fríða told me in a matter of fact manner that Ragnar's point was quite clear: "What is your real question here? I do not see your question?" she queried me. "He [Ragnar] tells you this, yes? But it is clear? Here, you know, yes, the glacier knows you are there, knows you are everywhere. I feel the glacier's awareness, and when I am walking in the mountains each [glacier] will always watch over me as I walk. This you always know living here."

"But, can a glacier watch over a specific person?" I asked her.

"Of course. It is watching over me, or her, as he [Ragnar] said."

Many Icelanders expressed to me perspectives parallel to Ragnar's that the glaciers were watching them—or their loved ones. A young woman who had lost her child told me, amidst personal distress still visible twenty years later, that she'd memorialized her infant daughter at her family farm in view of the ice. She told me she found comfort in knowing that the glaciers were watching her child rest now.

Speaking with another woman, Ásta, a ranger for the national park, she told me: "I used to sometimes, when we used to do guided walks around the glacier, sometimes I would take people around, and when it is foggy it is nice to sometimes take people to different places. And, even though you cannot see the glacier you can feel that it is there. It knows you are there."

Other people told me about times they felt the glacier was staring at them while they worked or visited areas near the ice. A young man in his late 20s, Sveinbjörn, told me about a reindeer hunting trip he had taken the previous year up by a local glacier. "Ya, that kind of struck me," he described. "The glacier was probably twenty or thirty meters or something, it was that close that I could spot it and, I felt it was really really weird. And I remember the guide, the reindeer guide, talking about the glacier there watching a lot, but I didn't pay any attention to it then. But when I came and saw the glacier, it was kinda weird to see how true it was. Seeing it with your own eyes was kinda of weird. The glacier was watching me."

OVER A POT OF COFFEE in January 2016, I spoke with Rúnar, an Icelandic artist living in Malmö, Sweden, who grew up in

Hornafjörður. While Rúnar had lived in Sweden since 2011, he returned to Iceland each summer to work in the region.

Rúnar spoke to me at length about how he thought there was a certain *aliveness* to glaciers. "I think it is also their livingness. To me personally, it is this idea of living, like the glacier is expanding, it's always moving. If you are in Skaftafell, you can hear the cracking and the sounds sort of like thunder. You can be in the middle of that valley and you can hear boom-shhhh, it always takes me a moment, oh! Glacier!" Rúnar explained. "There is something about them, they're not mountains because they aren't just like there, but also they are doing stuff, and to me, it is a bit like in a way, mysterious."

Rúnar was concerned for the future of that *livingness* of the glacier. Showing me a picture from a decade before, he told me a story of how when he was younger, he used to gather the trash from the area and burn it in front of a local glacier.

"What?" I responded.

"Yes!" he laughed. "I think glaciers are viewed a little bit differently now in a way, like this thing of burning garbage in front of it is kind of unthinkable today," he said. "I think back then glaciers were more understood but less respected in a way. I don't think people, I get the feeling that people in this area thought of them as a permanent thing, and now we're learning they probably are not."

Rúnar paused for a minute, then continued his thought to completion. "We watch them dying . . . and they are not living so much anymore."

To him, the glacier was living—and had been living—but it was dying now.

Articulations of glacier aliveness and dying were repeated to me over and over by many local Icelanders. For example, Jakobína, a scientist who worked with the Iceland Glaciological

Society (IGS), told me: "What I love about glaciers is that they are so alive. They constantly change. You start to, you know, you think you remember what a route was like and then you come back and it's so completely change. It's the thing that fascinated me so about them that glaciers are so much a living thing!"

Dúfa, an artist who lived in the region, told me she could not go back to her family's farm near Skálafellsjökull because "the glacier has died. I saw it grow once, and now it is dead. The blue is not blue, so I cannot paint what it used to be because it no longer is what it used to be."

A local man in his fifties, Ástráður, told me casually one day that he was quite concerned about the local glacier Lambatungajökull and how he "thinks it died. I was hoping, with the snow, that it would at least grow a little bit, but I doubt it now that I have seen it. The glacier will be less than last year. It will be even more dead than last time, the last time I came to the glacier."

Another man, Guðjón, characterized the glaciers of Hornafjörður in a mournful way. "We are seeing the glaciers pass away. That is a problem, a struggle . . . I think we are coming to the point where we have to make a stand. I believe that if we have this conversation as a whole community of Hornafjörður, we can help them slow their dying."

. N .
V ✦ A
· S ·

IN JANUARY 2016, my friends Hulda and Dögg and I drove out of Höfn as the sun rose around 9:00 a.m. We arrived at the turnoff for Heinabergsjökull near 9:45 a.m. just in time for the low-angled light to inflame the frozen groundscape with an incandescent glow. Neither of my friends had visited the glaciers in years. Driving with them through the area once we turned off the paved road was informative: in their eyes the land had changed; the Piedmont glaciers withdrawn like giant cat's paws retracting.

After parking in the car park for both glaciers, we walked towards Skálafellsjökull. I asked the women questions, and they answered eagerly, sharing one glacier story after another. Dögg described to me how winter was her favorite time to see glaciers. She explained: "In the midwinter, glaciers bloom. They are the clearest blue and shine at each other. You can see what they are thinking. In the spring, they turn white, they hide. I don't like them in summer, because then they are black, they are dirty and hidden and they cannot speak. This is life. But then, in autumn we have the rains here, and they are bathed and washed, and the bad skin is sanded away, and then they are smooth and clear and ready for snow dressing. This is the glacier life."

The cycle of glacier blooming—like a flower, like a life.

Glaciers bloom. In the spring, as the surface of the glacier is exposed to the atmosphere, the surface ice gets more "airy" and white. Through the summer, as the glacier surface melts, the sediments suspended in the ice—volcanic materials, atmospheric dust, etc.— are exposed and transported about through surface meltwater. Hence, the surface of the ice appears dark, dirty. But then comes autumn and rain and snow and the glacier is washed clean. The winter cold temperatures make that rainwater freeze on top of the glacial ice, enhancing the deep blue of the ice within.

Glaciers bloom blue through the seasons.

I asked Dögg to describe this cycle to me over and over. She did, willingly, reciting it like music.

. N .
V-✦-A
· S ·

ON THE WALKING PATH in front of my house in Höfn, I would sit on the bench by the water's edge and stare at Hoffellsjökull, Fláajökull, Heinabergsjökull, and Skalafellsjökull. I would greet local residents as they walked, jogged, strolled, or ambled by.

On days when the weather was clear and the glaciers were visible or "out," I saw many townspeople walk by my bench and stop. They'd pause, then turn away from my bench and face the ice. On particularly beautiful days, it was not uncommon to see ten or twelve tourists and local people standing still looking at the ice.

Sometimes I'd chat with people while they stopped and looked at the glaciers. One little boy who I met regularly on the town's walking path told me he always liked to stop and wave to the glaciers when they were out. He told me he liked it that they always waved back to him.

Perhaps thinking about a glacier's aliveness is to think about living, about what living means, about how we respond to the livingness of the world around us. In many ways, thinking about glaciers is also thinking about us.

. N .
V-✦-A
· S ·

JÖKULL, SOFFÍA, RÚNAR, and the others—and all the other people I talked with who are not quoted in these chapters—while at times slightly reticent about the potential livingness of ice, once the topic was opened spoke matter-of-factly about sentient glaciers. What I found fascinating was that many people spoke about sentient glaciers in such ways as to reveal a certain duality of knowledge.

Let me explain.

Ragnar perceived the glacier was alive—he *knew* it was alive—but the idea of an alive glacier did not dismiss the stories he had told me previously, stories that perhaps did not support sentient ice. The point of tension for him was never that such stories might be incompatible. Rather, it was communicating the glacier was untrustworthy. "You can see it, you can feel it, but you step up on the glacier, it is moving… I say to people, never trust the ice. Not completely," he said.

Communicating that the ice was alive was secondary to Ragnar's important point: never trust the ice. Ragnar was not implying the glacier's benevolence or malevolence; rather, he was conveying how glaciers entangled within the living landscape he contended with in his daily life. As his multiple stories of men falling into the ice, sheep going missing in the ice, glaciers receding, unsafe adventurers and such demonstrated, to Ragnar the ice was a strong, unpredictable force, shaper of fates, in the area. It watched. It cared—for Ragnar, for his family, for the community. The more Ragnar interacted with the ice, the more varied and conflicting and diverse stories he carried, and the more care he possessed for the glacier's livingness.

A different example: many of those who voiced perspectives of glacier sentience were scientists—people who might (stereotypically) not be expected to perceive life in glaciers. Broadly, those trained in a Western science knowledge framework[165, 166] have some difficulties supporting knowledges traditionally deemed outside that framework. Animism, perceptions of sentience in "nonhuman" things, etc. are tricky to prove—and evidence-based proof is one of the foundations of Western science.

Jakobína, the traditionally-trained IGS scientist, told me "they are so much a living thing, glaciers." While much of our conversation was also about how many glaciers were experiencing irreversible mass loss, to her, the glacier was also alive. These two knowledges, one aligned with Western science centering on ice mass loss, and the other on the perspective of ice alive, were not in conflict.

Jökull was a trained scientist, but he grew up in a culture where the landscape was alive. This was part of his social fabric. Having experienced Skálafellsjökull—his first living glacier—validated the cultural stories he grew up with, and influenced his ability to share the story of how Skálafellsjökull snarled at him, a moment which led him to develop a personal and intimate relationship with Skálafellsjökull. At the same time, he talked to me extensively about the glacier's more extensive physical properties and characteristics.

Soffía, the Icelandic scholar who just earned a doctorate degree, said of a glacier: "this was so old and this is so wise. It knows, this phenomena knows the stories of thousands of years." Soffía expressed that she felt the ice was aware, that it had been aware for a thousand years. But she said she also understood that "glaciers characterize the landscape here. . . . Iceland has an extreme landscape with the glacier and the geothermal areas, which make the whole earth, you know, visible to the processes shaping it." She spoke confidently about both the sentience of ice and the mechanics of ice upon the landscape.

When I asked other trained glaciologists in Iceland *about* glaciers—as in, I once asked local scientist Markús what Skálafellsjökull was doing that day, and he told me "he's fighting with the mountains," or, when I spoke with glaciologist Helgi Björnsson about the increasing number of tourists on Vatnajökull, and he told me the glacier "should not get mad at all the people,"–many scientists treated the glaciers as sentient creatures. In those moments, glaciers were alive.

Glaciers support dualities, contradictions, multiple knowledge systems. Glaciers partner with people, individuals, communities to enhance diverse ways of knowing.

. N .
V ✦ A
· S ·

TO THINK MORE ABOUT SENTIENT GLACIERS, it might be helpful for a moment to think about mushrooms.

Over the last decade, anthropologist Anna Tsing examined the matsutake mushroom to understand relationships between capitalism, collaborative human and more than human life, and multi-species landscapes.[107, 167-170] Tsing argued that the Western scientific "unit" designation (for example, specific species of trees such as Pine or Doug Fir, or types of forests such as boreal or temperate) can be limiting, dependent upon what stories one might want to tell about specific environments. If one wants to understand matsutake, she wrote, the unit is not the mushroom itself, but the complex assemblage (read: relationships) of matsutake and pine and oak forests, continental shifts, glaciations, soil, human foresters, mushroom foragers, and many others.[107] The story is not the mushroom alone, but rather, the mushroom and all the various relationships connected to the mushroom.

Tsing argued that the idea of a stand-alone unit of a species as "self-contained, self-organized, and removed from history," while fundamental to modern science, is a myth.[107] Many sentient organisms ("units") develop through relationships and partnerships with other organisms *and* things, including squids, wasps, butterflies, and ants. They would not be able to exist alone as a single unit, as a single species of wasp or glacier. They need the partnership to survive.

Extending Tsing's argument to glaciers, then, perhaps if one wanted to understand glaciers, then the unit of consideration is not the glacier body itself, but rather the complex relationships amongst glaciers and landscapes and people and missing family members and shades of blue and floods and so much more. Perhaps glaciers are really living constellations, part of communities of partnerships and relationships and knowledges that work together to make up the secret lives of glaciers.

Think back to chapter four, and concepts of assemblage. How certain things in certain arrangements create and/or reveal certain structures and relationships. Tsing writes, "if one wants to understand matsutake, the unit is not the mushroom itself, but the complex assemblage of matsutake and…"[107]

The "and" is crucial, likely the most important part of any assemblage, any relationship, any partnership.

Glaciers and.

That "and" might be the most important part of understanding perspectives of living glaciers. That "and" invites us all in, includes all of us, you, me, to partner with ice in our shared communities. It is not the glacier alone—just Skálafellsjökull out there potentially living and watching—it is what comes with—implicates—a living Skálafellsjökull and you and me and all of us.

Because here is the thing. If Skálafellsjökull is living, and Skálafellsjökull's "and" includes the people around the glacier and the communities sharing the surrounding environment, then at some point this might mean something to someone. Said another way, if we're all in this together, we might have to care for each other.

With fluctuating levels of explicitness, Ragnar, Rúnar, Ásta, Jakobína, Jökull and many others expressed care for individual glaciers and for glaciers as a whole.

Ástráður said he thought, "it died. I was hoping, with the snow, that it would at least grow a little bit . . ."

Guðjón believed that "we are coming to the point where we have to make a stand . . ."

Many different Icelanders expressed neighborly care for and of "their" local glaciers. Implied is they (read: people and ice) are in this together—mutually caring for the other. Soffía did not say she grew up in a world that had glaciers, rather, she said, "I grew up *with* that glacier." Like one might say, I grew up with my sister. With. Together. Soffía implied development and growth together. They grew together, her and the glacier. Care for sentient glaciers raises the possibility of genuine exchange between people and ice.[161]

While perspectives might vary, what doesn't appear to vary is that perspectives of sentience go hand-in-hand with expression of caring for glaciers. When glaciers are perceived as sentient, people care for them. And some people feel cared for by them.

Ragnar's last comment, though ambiguous, expressed his sentiments that ice watched over his deceased daughter. Several others articulated that the glacier watched over or protected them. Typically, when people ascribe the role of watcher over another human being, implied is some level of care. People watch over ones who are loved or cared about. The glacier watches over Ragnar's daughter, and he expressed comfort in this.

A connection between Ragnar and glaciers existed beyond the physical world and into the unknown—similar to the relationship Soffía referenced when she alluded to her childhood. She perceived that the glacier grew up as part of her childhood, a participant that changed as she changed. For both Soffía and Ragnar, the intangibility of ice and people extended into their own lives, included mutual, two-way care.

This is crucial: mutual care. By perceiving glaciers both as sentient and as living things to care about, such relations do not remain one-directional. People may care about a glacier, and the glacier may care about them.

The former President of Iceland, Ólafur Ragnar Grímsson, remarked that, "We have culturally, historically and politically, in all nations, been brought up with a view of Mother Earth in which the ice is peripheral. We have not acknowledged that in fact we all live in an ice-dependent world. Our weather, our climate, our crops and our cities are dependent, in one way or another, on what happens to the ice. The glaciers are not divorced from our fate; they are at the core of our future."[171]

"We all live in an ice-dependent world," President Grímsson observed. In Iceland, glaciers are as much dependent on people as people are of them, and this extends into notions of being human in this world. To care for glaciers, in this sense, includes caring for glaciers as independent of people, but also as connected to people. What happens to one impacts the other. There is then a mutual implied responsibility—perhaps even implied moral choices, issues of equity, or ethics.

Sitting on that ledge high above Skálafellsjökull, Jökull told me he wanted to take care of the glacier like he took care of his family, his super jeep, his home. I observed no ranking in the things he cared about—no one thing was more important than the other. All were within his purview of things he cared about.

Later that day, Jökull and I hiked back to the truck, which was still parked in the middle of the road, and we drove back down the dirt track from the ice to the main road. Before I hopped out of the jeep and into my car, Jökull handed me his phone and showed me several photographs he took of Skálafellsjökull.

He told me he was going to text them to his wife, because she was curious how Skálafellsjökull was doing that day. Jökull's relationship with Skálafellsjökull had expanded over the years to include his wife, his family, his life.

Julie Cruikshank's original question—do glaciers listen?—is still largely unanswered, but in a living sense, some might say yes.

I wonder if we are listening?

Perhaps with glaciers it is time to extend our collective imaginary rather than limiting it.[172] What might the world look like from Skálafellsjökull's perspective? What does being Skálafellsjökull *feel* like? What sensations does weather activate for the glacier? What is it like to burrow down into pliable bedrock? What does Skálafellsjökull remember, or hope? Who does Skálafellsjökull relate to: the mountains, the sea, clouds, you? Does Skálafellsjökull understand extreme ice mass loss? Can Skálafellsjökull still breathe? Does the moss growing on Skálafellsjökull's surface tickle? Does Skálafellsjökull listen?

CHAPTER SEVEN

≈ ≈

"I think it is a time bomb, right now.
This winter is the largest one in our ice cave trips."
ELÍN, 2015

"Everybody is trying to profit off the glacier."
GUNNLAUGUR, 2015

"I'm not going to spend my life waking up around 6 o'clock going to milk cows and go to sleep around midnight every day, always the same thing. And my parents they are doing this very well and they are getting old and I don't want to do that."
BRYNDÍS, 2016

I HUDDLED DOWN NEXT TO the front wheel of my friend Snorri's super jeep and tried to affix pull-on front spikes over my alpine boots without taking off my mittens.

The weather was February—cold, billowy, with a storm slamming the south coast, grounding all domestic flights and closing roads in southern Iceland from Hvolsvöllur to Breiðamerkursandur, the large glacial outwash plain in front of Iceland's third largest outlet glacier, Breiðamerkurjökull.

I was actually standing *on* Breiðamerkurjökull, about to walk into one of the glacier's two large ice caves. I was with Snorri and his foreign client, Jack. Because of the storm and road closures, no other guide companies were taking tourists out yet, and we had the remarkable opportunity of being inside the Crystal Ice Cave completely by ourselves.

That trip was my twelfth. By March 2016, I'd have visited the Crystal Ice Cave fifteen times alongside twelve additional trips to other area ice caves. The ice cave season in Iceland typically runs from November to March, with guides and tour operators waiting for temperatures to drop low enough to stabilize before safely offering day and night tours.

I visited ice caves throughout the winter as I tried to understand people and ice on the southeastern coast. I accompanied guides and tour operators while they worked with clients, helped with tasks including glacier education, safety assistance, repairing infrastructure such as bridges or fixed lines, and traveled on several "sussing-out" trips to check out potential new caves or gauge an area's accessibility and safety. The glacierized landscape changed so often (hourly and daily) that guides from different companies cooperated to check out and report back conditions, changes, and hazards.

Ice caves are exceptionally dangerous and should not be accessed without certified guides or specialized knowledge. Visitors to any ice caves should utilize local guides. Mitigating the risks is a guide's full-time job, what one guide characterized to me as like a "spider with all my antenna up and watching, many eyes and ears out all watching." Safe access regularly requires modified vehicles and equipment including crampons, full body harnesses, safety gear, and ice tools.

"Ice caves" reference hollow, accessible features within glaciers—typically moulins, subglacial tunnels, or channels that cut through the interior of a glacier. Iceland has many, many ice caves throughout the island's many glaciers, but in the winter of 2015-2016, there were only two available by guide on the south coast—both located within the glacier Breiðamerkurjökull. The ice caves were known as the Crystal Ice Cave and the Waterfall Ice Cave. I describe these ice caves throughout the book, though I do not provide exact locations for any. Given the recent tourism increase in Iceland, many tourists have tried to access the ice caves without purchasing guides and have subsequently exposed themselves to extreme hazards.

Snorri, Jack, and I worked quickly, pulling on gear and stacking tri-pods over backpacks. The four-minute traverse from the jeep to the entrance of the ice cave was like walking through a wind tunnel blasting snow, ice, and sand.

But instantly, upon entering down *into* the glacier, *into* the ice cave, everything settled calm and silent. The wind, weather, and cold vanished, and we walked inside the crystal-clear, Picasso-blue ice cave antechamber full of hazy, sapphire-toned Arctic light.

Snorri smiled. Jack was speechless.

We were coated in blue blue blue light.

Winter after winter for the last five years or so, drawn by the extraordinary blue ice caves, thousands of people have traveled to southeastern Iceland. The winter of 2015- 2016, around 60,000 people alone purchased guided tours to ice caves in Breiðamerkurjökull. Unquestionably, the ice caves create considerable draw for winter tourists, and are, to quote multiple local guides, "phenomenally lucrative."

But I found that a curious tension materialized within the coveted blue light at Breiðamerkurjökull—a tension between loss and gain that many local people did not know quite how to reconcile. Some of the most obvious symptoms of global climatic changes in Hornafjörður are unprecedented glacier loss, and many people living in the region expressed concerns about the recession of the area's ice. Concerns were articulated in many ways, including, for example, a local artist, Rúnar, who said, "Sadly I think that they are going to disappear, and it is also kind of unthinkable."

A local teacher, Jensína, told me, "In my childhood, my glacier, it was something that my grandmother loved. She'd lived there near that glacier her whole life. She just loved that glacier. It was her glacier. I was just a child when she told me all the names of the glaciers. You know, it has always been there, I cannot think of it gone now."

And a local environmentalist, Hjörtur, confessed, "I am having difficulties reading articles and news, seeing this about the glaciers. It breaks my heart, actually. I think this is really sad. And sometimes I just look the other way, because I feel like there is nothing I can do about the glacier." Young people under the age of thirty especially expressed concerns about glacier recession.

However, at the same time people spoke of their concerns for glacier loss, many also articulated an unmistakable awareness that, at least over the short-term, glacier change was beneficial to the people in the area.

Tourism in Iceland has grown exceptionally fast, with a 260% increase in tourists between 2010 and 2016. In 2016, 1.76 million tourists traveled to Iceland, and about half a million of them traveled to the southeastern coast attracted by glaciers and glacier-related activities.[24] And while in previous years glacier-related tourism was relegated to the island's short summer months, now, with the ice caves, tourists come year-round to experience glaciers in summer or winter capacity.

The impacts of rising year-round tourism over the last couple of years are profound in Hornafjörður, especially as the region has been historically poor and remote with few natural resources and a long tradition of being left isolated to its own devices.[3, 56] Then, what slow economic progress the area had made, particularly through the 1990s, was diminished by the impacts of the country's 2008 financial crisis (often referred to simply as "the Collapse"). The country, and especially Hornafjörður, were quite slow to recover after all three of the country's major banks defaulted and created the largest systemic banking collapse in the economic history of the world. Iceland plunged into an economic depression, the Icelandic króna plummeted against the euro, inflation skyrocketed, unemployment tripled, and many Icelanders lost their homes, jobs, and pensions.[173]

This is the backdrop. Now, overlay over the last five years a situation where locals of Hornafjörður start to acquire year-round economic and social benefits through year-round tourism—largely centered on glaciers. But it isn't just that tourists are attracted to glaciers. Tourists are also attracted to glaciers because they're influenced by what tourism scholars characterize as "last chance tourism," where travelers flock to see destinations perceived to be or soon to be devastated by climate change.[174-177]

In Iceland, tourists come to see the glaciers before they're gone.

In just the last few years, many Icelanders have regained post-Collapse financial security through self-determined employment guiding, opening year-round hotels or restaurants, or other periphery forms of occupation as the economic ripples of glacier-related tourism reverberate through the region.

One owner of a Reykjavík-based tourism company, Höður, summed up the situation: "As the glacier melts, it melts money. We will sell each year more tours to ice caves, and I know more ice caves will open with better temperatures. Now I need more ice caves to open up closer to Reykjavík, but we can make some then if it does not do it next year or after then."

What Höður meant is that he foresaw the ice cave industry growing, and that he hoped ice caves would open within glaciers closer to the country's capital with more people, more tourists, and more money to be made.

Because, as he explained, as the glaciers melt, they melt money.

Here then is the crux—and another element of people-ice relationships on in Iceland. Glacier-related tourism (especially ice cave tourism) and climate-change-exacerbated glacier losses are coalescing, straining people and ice in increasingly complex and contradictory ways.

The complexity of short-term positive and negative benefits of environmental change are inherently at odds with larger perceptions of climatic changes as negative or catastrophic.[8, 178-181] To be clear, I do not deny the seriousness of climatic changes, nor their anthropogenic nature, nor the unassailable reality that Iceland is indeed losing ice at rates never previously seen, a fact indisputable and supported by substantial evidence.[3, 40, 41]

What I am highlighting here is that many people on the southeastern coast are currently benefiting from glacier loss—and by extension, climate change—and a narrow focus on only the negative aspects of glacier change risks reduces glacier change to a single story of loss and tragedy that fails to represent, understand, or acknowledge the harrowing complexity and immense range—positive and negative and every varied experience somewhere in between—of lived people-ice experiences and interrelations across the world.

If we are told there is only one story of change, and it does not match our own lived experiences, then we are removed from the overall story. It is vital to understand the on-ground experiences of climate change, and question what benefit we actually gain by parameterizing climate change within a value-based conversation of positive or negative. I suspect that value-based conversations of change are too porous, too subjective, too individual in nature. How I value something might not necessarily match how you value the same thing. Looking at what is happening on the southeastern coast, it is clear we need to re-think how we frame our conversations on change. What I am advocating for is to frame climatic changes with a context of mutual transformation—where all of us are transform in concert with our environments.

Let me explain.

. N .
V ✦ A
· S ·

THAT MORNING, Snorri and I had picked up Jack at the N1 fuel station on the edge of Höfn on Vesterbraut where Jack had spent the night in his car—with the recent high volume of winter tourists visiting the south coast for ice caves and northern lights, accommodation had not caught up with demand. Many nights, the 2,000 beds available in Höfn that winter were completely booked.

Jack was none the worse for wear, so we were quick to pick him up and get underway. Snorri kept an eye on the storm, and I entertained Jack with glacier-related science. In under an hour, we arrived at the turn-off to Breiðamerkurjökull.

Breiðamerkurjökull is the second largest glacier outlet draining off Vatnajökull, and the third largest outlet glacier in Iceland. The glacier flows approximately thirty miles off the ice cap and has a terminus face over nine miles wide.[182] Like most glaciers on the south coast, Breiðamerkurjökull started to appreciably recede around 1890. It retracted over four miles in the twentieth century alone, and gouged out a series of proglacial lakes, one of which is the deepest (and growing) proglacial lake in Iceland: Jökulsárlón.

The turnoff was a dirt track that traced a path out across Breiðamerkurjökull's glacial outwash plain, Breiðamerkursandur. Snorri drove his jeep slowly. The snow was pouring down, and the road was covered. The road was white, the sandur was white, the sky was white.

Even going slow, twice the jeep slid off narrow plateaus and punched down into depressions full of snow that required over fifty minutes of digging with shovels in order to break free. Jack seemed to find the whole experience quite adventurous, and Snorri just patiently dug out his jeep again and again and again.

Such was driving on a sandur created by Vatnajökull's second largest outlet glacier, a landscape comprised of moraines, drumlins, eskers and other glacial geomorphological features that are all rapidly transforming.

Locals spoke often of the startling rate of Breiðamerkurjökull's recession. For example, Vök, a thirty-year member of the Icelandic Glaciological Society, observed to me, "Like what has been said many times, Breiðamerkurjökull is moving now incredibly fast, leaving behind a trail of gravel and drowned lagoons."

Exact recession rates have been difficult to assess, as the glacier's three largest proglacial lakes, Breiðárlón, Jökulsárlón, and Stemmulón, each exhibit different behaviors. Between 2000 and 2009, Jökulsárlón increased from approximately six square miles to eight square miles, a change not matched by Breiðárlón nor Stemmulón—until, that is, Stemmulón recently merged with Jökulsárlón.[183]

To complicate matters, Breiðamerkurjökull itself encompasses over *fourteen* distinct streams of ice within *three* major ice lobes. Just as each proglacial lake behaves differently, so too does each ice lobe. The eastern and western lobes flow across bedrock, but the larger central lobe flows partly over bedrock and partly over seawater within Jökulsárlón. This central lobe arm drags in ice flows from either side, visibly distorting neighboring flows, moraines, and internal morphologies.

Not all of Breiðamerkurjökull's recession is attributable to air temperature increases brought on by climate change. At least for the portion of the glacier margin terminating at Jökulsárlón, local dynamics interacting amongst the glacier's calving front and warm seawater exacerbate the rate of recession.[3] Some scientists suggest that once Breiðamerkurjökull's recession increases, the glacier in the next fifty to one hundred years might recede all the way to the nunataks of Esjufjöll and Mávabyggðir.[3, 183]

Once our jeep was off the sandur and driving atop the snow-encrusted glacier, we stopped getting stuck and were able to move more quickly. In just a few short minutes, Snorri got us to as close to the ice cave as possible. The weather was pitching, blasting snow and ice, so we organized our gear as much as we could before bailing out of the jeep and dashing for the cave.

Inside, Jack had his camera out in less than five seconds. He was ecstatic. Today, the ice caves in Iceland are among the most photographed places in the country, pictures of unbelievable hues of sapphire, bubble-waved walls of sheer blue resembling underwater worlds, of time-stopped frozen waterfalls, ice tunnels, ice stalagmites, incandescent sublimation crystals and jagged icicles, low winter light pouring through blue ice caves filtering flickering blue-golden glows.

It is truly one of the few things in the world that people must see first-hand to believe.

. N .
V-⊕-A
· S ·

A PREVAILING ASSUMPTION about glacier change made by academics, scientists, analysts and many others—and reflected in public discourse—is that glacier loss is an inherently negative experience for peoples and communities. For example, former US Vice President Al Gore wrote that if there is no collective action to "…slow and gradually halt the current meltdown, we risk destroying the very global systems that have enabled us to thrive and prosper."[27]

Media headlines frequently proclaim, for example, "White House warned on imminent Arctic ice death spiral,"[184] "The melting of Antarctica was already really bad. It just got worse,"[185] "Polar melt down see us on an icy road to disaster,"[186] and "As Greenland melts, this iconic glacier is creating terrifying tsunamis."[187]

As noted at the beginning of this book, glacier change has been described as parallel to New Yorkers' experiences with the 9/11 terrorist attacks. As one scholar wrote, "There are cultural impacts of glacier retreat. Many human societies have strong attachments to glaciers… These features have strong symbolic significance, and people identify with them… If a rather extreme parallel may be drawn, many people experienced deep distress over the attacks on the World Trade Center in New York not merely because thousands of people were killed, not merely because valuable property was destroyed, but also because of the symbolic importance of the buildings themselves…"[7]

Glacier change is routinely associated with sea level rise, fluctuating hydropower generation capacity, conflict over water resources, increasing hazards, and cultural degradation, among other issues.[7, 85, 146, 188-190]

Such issues to some degree find accuracy on the ground. In one study, researchers reported notable levels of anxiety and fear surrounding glacier recession in a comparative anthropological study in the mountain ranges of Cordillera Blanca, South Tirol, and North Cascades.[85] Locals from all three areas articulated negative views of "deteriorating" glaciers posing a "major challenge" to communities; fears regarding future water scarcity, aesthetics, decreasing mountain tourism, and decreasing water quality; and feelings of helplessness, sadness, and shock regarding the rate of glacier change.[85]

Overall, the framing of glacier change is negative—a framing not entirely undeserved. Research has demonstrated the many inarguably *negative* impacts of glacier loss. As environmental historian Mark Carey writes, "[n]atural disasters, water shortages, degraded tourist destinations, and vanishing mountaineering terrain—these are among the ill effects that melting glaciers and global warming increasingly produce around the world."[17]

However, broad strokes tend to overlook the complexity of daily, lived experiences between people and ice. Just because glacier change registers negative impacts in some places and some times does not mean it holds true for *all* places and *all* times.

Consider, for example, as earlier chapters described, how some older Icelanders in Hornafjörður feel immense relief, safety, and security that ice will no longer destroy their homes or livelihoods as it repeatedly did over the last thousand years. Or, how some Icelanders are creating new jobs and short-term financial security in glacier-related tourism. Or, how some younger Icelanders no longer have to leave the region after schooling for work in the city. Instead, they might find year-round, well-paid work on the coast. Or, how some entire communities no longer face unpredictable glacial outburst floods, or how local farmers may now access to new, fertile, ice-free landscapes, or how increases in short-term energy production occur with more water melting into the country's hydroelectric power generation system.

Glacier change is not *all* bad. It is also, I stress, not *all* good.

Change is never all good or all bad. And all environmental change is a mixed bag, based on time, scale, geography, and the complexity of ever-shifting human points of view. In fact, many scholars have recently begun arguing not only against a singular negative focus on glacier change, but also against a narrow negative emphasis in climate change discourse.[8, 178-181, 191]

For example, anthropologist Kirsten Hastrup argued that "climate is no longer seen to make places but rather mostly to destroy them… the very minute that 'climate change' is invoked and studied, cultures are already perforated… casting this in terms of cultural loss may be counterproductive, quite apart from being an inadmissible confinement of people to particular and rather closed worlds."[191]

Geographer Mike Hulme warned against reducing complex environmental changes down to single stories of loss and tragedy,[178] and geographers Karen O'Brien and Robin Leichenko argued that assigning claims of value for climate change impacts produced perceptions of climate change 'winners' and 'losers.' However, they found that the claims held little actual merit as they were typically general in nature, did not account for dynamic short and long-term time scales, and were dependent on a host of multi-scalar social, economic, political, and environmental variables.[192] Critically, they observed a greater focus in scholarship on people and communities labeled "losing" at climate change, and a reluctance to identify (or even acknowledge) perceived climate change "winners."[192]

The human experiences of glacier change—any change—is

influenced by how that experience is valued. How people understand change, and what value they assign it, filters through their own culture—and, I would hasten to add, their own personal experiences.

To me, walking up to a glacier I have known my whole life and seeing firsthand how much of that glacier has vanished—unquestionably this is a negative experience to me, a gut punch that leaves me gasping. But that does not mean the experience is the same for every person who sees that same glacier, or the person living next to that glacier that no longer has to worry about hazardous floods.

Longer timescales (many models of Iceland implement hundred-year time periods) suggest climatic changes will likely have far-reaching negative consequences in Iceland that could theoretically jeopardize the future of human habitation on the island, including the transformation of precipitation rates, increased air temperature, glacier loss and reduced glacier runoff capacity, isostatic rebound and correlated increased volcanic activity, strained flora and fauna ecosystems, warmer ocean temperatures, southerly shift in the Gulf Stream, and increased flooding and landslides.[193-197] All such changes will undoubtedly have complex impacts on social, political, economic, and environmental processes. But at present, the country is experiencing a long twilight where some Icelanders may savor some beneficial ripples of glacier loss. Undoubtedly, this moment will not last long, but it *is* part of the present, lived experience of people and ice in this region, which impacts then how people perceive climate change more broadly—and how they write themselves into greater stories at play across the planet.

To understand how people live in a world *with* ice—this very moment—it is essential to comprehend complexities beyond widely narrated negative consequences: the whole picture is needed, otherwise, the entire experience of glacier change is reduced to a single story of loss.

What is needed is to understand how peoples everywhere live with transforming ice—and how they themselves are transformed alongside the ice.

A last important note. A climate denialist or obstructionist might look at the evidence contained in this chapter and incorrectly conclude that climate change is good, or that climate change creates better environments. They might argue that at present, Iceland's temperatures are growing warmer, diverse fish are showing up in local waters, rivers are more manageable, hay crops are doubling, vegetables are thriving, and tourism is mounting—all positives changes in certain senses. But while glacier recession has increased the financial gain of many, it has also brought increased conflict. Glacier recession has increased senses of ownership of local ice, but also increased an awareness of significant loss. Glacier recession has increased local exposures to global narratives, but also fractured local knowledge. I repeat: glacier change is neither entirely positive or negative. Glacier change is transformation.

TOURISM IN ICELAND was largely perceived as "manageable" throughout the 1990s and into the early 2000s. As many Icelanders can recite by heart, tourism accelerated after 2010 due to two serious disruptions: the financial collapse and the eruption of Eyjafjallajökull.[198] Before these events, Iceland held a reputation as an expensive travel destination, which in many ways stymied tourism.

However, the 2008 financial collapse transformed Iceland overnight into a place affordable for middle and lower income tourists as the national currency plummeted. Two years later, in 2010, the volcano Eyjafjallajökull erupted, grounding millions of air passengers worldwide for over six days. Media descended on the island (the Icelandic airport remained open during the eruption, of course) to cover the volcanic eruption and its impact on travelers.

Images of Iceland's landscape—glaciers, volcanoes, geysers, etc.—were projected into televisions and computers worldwide, and, by 2011, as one local woman observed, "we were overrun by humanity."

The number of tourists traveling to Iceland in just the last couple of years alone is astounding. In 2016, the Icelandic population was outnumbered by tourists roughly six to one. By comparison, in the same year, visitors to France outnumbered the French two to one.

Recall, Iceland has a population of approximately 350,000 people. In 2016, 6.7 million people arrived at Keflavik Airport, 1.76 million visited the capital, and approximately 439,000 traveled to the southeastern coast to visit Breiðamerkurjökull's Jökulsárlón.[199, 200]

Jökulsárlón is an exceptional place in Hornafjörður, and it is central to the region's summer and winter glacier tourism. When Breiðamerkurjökull began advancing around 1600, the glacier surged over and over until it reached its maximum extent in 1890, just 300 meters short of the North Atlantic Ocean. As the glacier started receding, it began carving out Jökulsárlón.

The North Atlantic Ocean and Jökulsárlón are connected by the island's shortest river, Jokulsá, which is just shy of a mile long. The river splices the narrow crust of land between the two bodies of water, and acts as a conduit for warm ocean tides to enter Jökulsárlón and interact with Breiðamerkurjökull's snout face. Relatively, ocean water is much warmer than ice (with its lower albedo and higher heat capacity), and as such the warmer waters exacerbate Breiðamerkurjökull's melt rate and trigger the profuse calving of icebergs—all of which tidally circulate within the proglacial lake and attract thousands and thousands and thousands of tourists.

Breiðamerkurjökull and Jökulsárlón are now year-round tourist destinations, which has caused tremendous problems for the management of Jökulsárlón and the surrounding landscapes, many of which are outside the scope of this book. And while tourism used to be confined to summer, with the new addition of ice caves as an attraction, now the tourism season rarely slows.

Tens of thousands of people come now each winter to the tiny, two-bathroom rest stop on the south coast to see the icebergs, the northern lights, and—as Jökulsárlón is the meeting point for many tours—meet their guides for ice cave tours within Breiðamerkurjökull.

More people visited Jökulsárlón in 2014 (~388,991) than the number of people living in the entire country.[201] By 2015, 20.6% of all winter tourists in Iceland visited Jökulsárlón, and in the summer, 42.3%.[200] Jökulsárlón receives more visitors per annum than any of the other glaciers lagoons discussed in this book by several orders of magnitude, and is an essential element within the 2.3 billion USD growing tourism industry in Iceland.[202] At the time of writing, the industry exhibits no signs of slowing.[202] While tourists report traveling to Iceland for many reasons, part of the draw is to see and/or experience glaciers, including visiting ice caves.

During the time of my fieldwork, Breiðamerkurjökull had two ice caves accessible to tourists: the Crystal Ice Cave and the Waterfall Ice Cave. I explored other ice caves within Breiðamerkurjökull, Skálafellsjökull, and Fláajökull, and visited caves to the north of Vatnajökull in Kverkjökull and within the island's smaller ice caps. More broadly, I have visited ice caves in Alaska, Washington State, Canada, Scandinavia, and many other places.

My point here is to point out the commonality of ice caves—the feature is quite usual even as it is marketed as rare to tourists. Several times I traveled with tour companies that took people to what could only be described as glacier margin overhangs. The tourists, not knowing any better, were thrilled. This distinction is important, because ice caves come in all shapes and sizes; as such, tourism operators can sell multitudes of different features as "ice caves."

Broadly, ice caves form through ablation, which is the process of ice mass loss through ongoing evaporation, sublimation, melt, and erosion catalyzed by environmental and human processes. Ablation is typically sorted into two different processes: physical or mechanical. Physical ablation includes snow and ice melting on the surface, interior, or bottom of a glacier, or through sublimation or evaporation of flowing meltwater, or ice melt upon contact with ocean or lake water, or through contact with geothermal activity of some kind. Mechanical ablation refers more generally to weather propelling snow from a glacier, or calving ice, or ice-filled jökulhlaups, or even snow and ice avalanches draining from steep alpine glaciers.[34] I suggest a third ablation category: human-ablation. While most often seen in the instances of mining, tourism has increased incentives to manipulate glaciers (read: make, stabilize, or drain ice caves) with heavy machinery to manipulate or "improve" ice caves.

What is important to focus on is that glaciers do not just lose volume from surfaces or termini; mass loss also occurs *inside* and *under* the ice—and as glaciers lose ice mass, hollows, depressions, or cavities may form. These new features are called "ice caves" in Iceland.

The two accessible ice caves that winter were very different experiences. The Crystal Ice Cave, west of Jökulsárlón, was a large, nostril-like hole near the terminus of the glacier's center lobe. Access involved driving over the rugged outwash plain in vehicles with high clearance, and then up onto the glacier Breiðamerkurjökull. The ice cave itself was a sizable water-carved conduit with a series of antechambers progressively smaller and smaller the further inwards the glacier one goes.

The Waterfall Ice Cave was quite different. Situated on the extreme eastern edge of Breiðamerkurjökull, the ice cave was right where the ice rubbed up against the side of the mountain Hellrafjall. The valley Veðurárdalur funneled down towards Breiðamerkurjökull's flank, gathering and channeling the area's water catchment into a multi-tiered waterfall that cascades directly onto the glacier. This mechanism is what bore out the ice cave.

Of the two, the Waterfall Ice Cave was much trickier to access, with tourists driven out in high clearance vehicles on an unmarked track, then required to clip into fixed lines to traverse the edge of the mountain, walk across a portion of glacier and scree slope, cross a river and the waterfall, then descend under the mountain and the overhanging lip of the glacier's flank into the ice cave. Towards the beginning of the ice cave season, guides from several local companies would make daily trips to either of the ice caves to report if water had slowed or froze; as soon as the caves "dried up," tourists could be brought in.

Ice-cave tourism should occur in the winter, when the processes that created the caves—moving water, evaporating ice—slow. In winter, ice caves are not static, they transform all year long, but in the winter, typically such processes are slowed. Some caves are stable—such as the two caves in Breiðamerkurjökull—but others are not, dependent upon a host of local conditions.

Both the Crystal Ice Cave and the Waterfall Ice Cave are located on periphery edges of Breiðamerkurjökull—within ice that is continuously moving and morphing. Local weather, such as heavy snows or rains, impacts the dynamic caves; several times over the winter Breiðamerkurjökull's caves were flooded due to heavy rain and became unsafe and inaccessible. The dynamism of ice caves is often at odds with schedules, tourist expectations, and the extreme financial pressure some companies apply to get as many paying tourists in and out of the ice caves.

Ice cave tourism, while to some still perceived as in its infancy in Iceland, has started to catch up to what was previously a summer-dominated glacier-based tourism trade on the south coast. Summer available activities traditionally include super jeep and snow machine tours, guided single and multi-day walks, hikes, and climbs, boating on proglacial lakes, photography tours, flightseeing, and dog sledding.[24, 203]

How ice cave tourism started in Iceland centers on Breiðamerkurjökull. The story repeated to me by various people living throughout the region always involved one local man, Einar Runár Sigurðsson.

Alfa, a local woman in her twenties explained, "Einar has been going into ice caves for a long time, for over twenty years. But he didn't think about bringing people there, he just went there by himself. And with his camera. And then he was getting requests, hey, can I go to ice cave with you?"

Einar considered it. He'd worked as a professional guide and photographer most of his life, and in his spare time, he photographed the surrounding nature, including ice caves. In the early 1990s, he opened a formal guiding company, and in 1994, he started offering trips summiting the tallest mountain in Iceland, Hvannadalshnúkur. The company employed his wife, son, and daughter-in-law, and after Einar's photographs went viral on the Internet, his company was the first to offer guided ice cave trips.

Alfa explained: "That's how it started. Enough pictures of Einar's started to spread around, and people saw them and wanted them and wanted to go there and so [the guide company] started getting more and more requests. [The guide company] was the only company doing the ice caves tours in the beginning and were offering one daily trip and it never sold out. And then, last season [2014], it went bam, and now this is like madness."

N
V A
S

SNORRI TOLD ME, as we stood within the Crystal Ice Cave and watched Jack literally jump with incredulity in the blue blue light, that he felt tired. Not of going to the ice caves, but physically tired, because he used to guide all summer and then rest in the winter.

But now, with the ice caves and winter tourism, he felt like he was guiding every day of the year. He explained that he "needed a break. Soon. I don't need to be rich, I just need to put food on the table. And pay for everything." But he couldn't take a break because the money with the ice caves was too good. "It pays for all of this," he explained.

Glaciers have material impacts on the resurgence of the region. Previously slumped, the post-Collapse economy today is thriving (albeit with some conflict and strain) in large part due to glacier-related year-round tourism. Peoples' lives in Hornafjörður are progressively entangling with ice as locals' glacier interactions intensify; Icelanders are securing financial stability (a luxury after the Collapse), a sense of ownership over local glaciers, and a means of returning home to the region.

But the situation is intricate. Locals are also trying to negotiate what it means to "benefit" in some way from climate change while also learning to navigate increased glacier-related conflict and strains on local systems. Climate change is never far away during these negotiations—especially from the tourists themselves.

In February of 2016, I spent several days at Jökulsárlón talking with tourists who were about to embark on, or had just returned from, an ice cave tour. I asked them about themselves, and why they did (or were going to do) an ice cave tour. Most tourists offered multiple reasons, including the rarity of the experience, the color blue, seeing pictures online, the beauty of the winter landscape, and a chance to see the glacier/ice cave before it was gone.

My findings align with the conclusions of other researchers who observed a growing trend of "last chance tourism"[204] regarding tourism and glaciers. Tourists travel to Iceland for a variety of reasons typically related to the country's nature and glaciers, and climate change is within those constellations of reasons. Some people just want to see glaciers before they disappear.

One local ice cave guide, Hreiðar, observed this trend repeatedly. "It has something to do with when the media talks about the global warming and the glaciers melting and people start to think, it's in the media and the glaciers are melting and the ice is going away! They [tourists] go, 'I need to go and see a glacier before it melts away.' Something like that. Then they come here and ask me if I believe in global warming."

Many guides reported having ongoing conversations with clients about climate change at different scales—what was happening locally to particular ice caves and glaciers, nationally to Iceland, and globally.

As Snorri and I stood in the ice cave watching Jack photograph the ice cave, he told me how "Tourists tell me about their homes, but they can't see climate change there. They come here to the glacier expecting to see climate change. I feel a little bit sad about that." Snorri suspected tourists came to Iceland, and to the ice caves, in part to *see* climate change.

In other words, tourists came to Iceland expecting to see change in the glaciers—which can be rather hard if you're not seeing the ice regularly. But because many people are trained to see glaciers and then read climate change, tourists would see the ice and subsequently "see" climate change.

. N .
V ✥ A
· S ·

DURING THE TIME OF MY FIELDWORK, there were seven local companies (local defined in this case as businesses physically based in Hornafjörður) offering ice cave tours, and at least ten larger companies based out of Reykjavík. Two new local companies formed over that winter, and several people I interviewed spoke of potentially opening up their own company in the next several years.

One man, Fannberg, a stay-at-home father, told me: "it's like a gold diggers' mentality going on now. Everybody just invest in a big jeep and makes a year's labor out of four months or something like that. I am not in, but I have been thinking about it, and I'm not sure if I will, because of the children."

Of all the glacier-related tourist activities, ice caves were by far the most lucrative. And the financial benefit of the ice caves was not isolated to guides or tour operators; notably, the entire Hornafjörður economic system appeared impacted by the increase in winter tourism.

Previously, while tourism was an important industry in Höfn, it was largely confined to the summer months. But in 2015, tourism overtook fishing as the town's primary industry—year-round.

Now, accommodations that previously would have been open just during the summer were extending through most of the winter. Of the region's sixty-four available hotels, guesthouses, hostels, apartments, and campgrounds, only twenty-two stayed closed for the entirety of winter. The other approximately 70% stayed open. This was a significant shift from just three years previous, in 2012, when the percentages were largely reversed. But even with more accommodations opening, many places were completely booked. As such, many local people began to make additional money renting out bedrooms, living rooms, and yards for campers through online services such as Airbnb.

In October of 2015, I met with Dadda and Dagfinnur, the couple with the young son from the fourth chapter. The couple operated a guesthouse in Höfn, but were about to start a guiding company to take people to the ice caves. Neither had backgrounds in tourism nor glaciers, but both were excited with the new opportunities presented by ice caves. The story of how they decided to start their company was analogous to other stories I heard from the other local companies.

Dadda recounted that in their guesthouse, tourists always wanted to know how to get to an ice cave: "the knowledge is very low about ice caves, so the tourists think they can just drive right up to the ice cave. No problem. And they ask me, where can they take this blue blue picture they've been seeing?"

Dadda said this happened over and over, year-round.

Dagfinnur picked up the story. "So, we went first into glacier hiking because I found that I was always calling [a local guide], I was always calling [a local company], calling these guys for the guesthouse because our guests wanted to go, and the companies were always either fully booked or the guides weren't going. So, that's okay. I decided I will just do this. Not hard. But many of the locals have stopped doing glacier hikes, most focus on ice cave tours. There is more money there. So we decided to do glacier hikes!"

The couple picked a name for their company in English after doing some Internet research. They wanted to associate the name of their company with glaciers to optimize business. This was standard—most of the guide companies in the area use the English word 'glacier' somewhere in their name.

As one young local man, Pétur, drily observed, "Even glaciers have glaciers in the title of their name. So now, glaciers guides, glacier walks, glacier jeeps, everything it is a part of the marketing, a glacier is part of the marketing of the area."

What Pétur refers to is how often in English one might see the glacier Breiðamerkurjökull written as Breiðamerkurjökull Glacier—which essentially reads as Breiðamerkur Glacier Glacier.

I periodically checked in with Dadda and Dagfinnur over the course of the winter, and by all accounts, they were doing well. They had, almost predictably, expanded into ice caves.

"Everyone wants to go to the ice caves, it is easy to do!" they told me.

Easy, while an understatement, is important to understand here. As opposed to glacier hiking, or other glacier-related activities, ice cave operators need less start-up investment, equipment, specialized knowledge, and staff. Bare bones guides could work for themselves and utilize their personal vehicles to transport people to ice caves. Equipment could be minimal—I witnessed many guides in the ice caves without safety equipment (helmets, ropes, harnesses, crampons or slip-spikes, ice tools, etc.). Conversely, I saw multiple instances of some local companies fully gearing up clients and taking them safely into the caves.

The point is that the start-up needed to open an ice cave-related tourism business was relatively low, and many local people entered the business in part because of its relative ease. As several people observed, if it was not so easy, what might have happened over the last two or three years was the creation of one or two companies, and then many locals would have worked for those few companies. Instead, while there are different sized companies, many people I spoke with operated independently (sole owner-operator) and contracted out with their fellow companies/neighbors if many guides were needed all at once for larger parties.

Snorri was a good example of this; he had his owner-operator company, but he was available to contract out with other local companies as needed. When he did, he brought a personal vehicle, safety gear, and equipment.

Locals did tend to invest once their companies got going. By the end of winter, Dagfinnur was driving a large multi-passenger van and had a fair amount of safety equipment and gear.

Another Höfn man, Ari, started his glacier business with his wife and friends in 2014. Ari explained this decision to me: "I was working for the municipality. You had actually endless obligations, and it didn't matter how much you worked, or how much effort you put in, it didn't solve any, what could you say, income problems. So, then one time, I decided to find something that I could trust myself to start and try out, and put all of my effort into that, instead of putting my effort into something that was maybe not very well appreciated and not very well paid. So I decided why not stand by my own and try something out for myself and see how far I can get."

Ari's company started with a single vehicle with six seats that went to the ice caves twice a day. For the first six months, Ari drove each day on his own. He also built the company's name and website, and the outfit grew from there. Once the company got off the ground, Ari and his co-owners invested significantly. By the end of the 2016 winter, the company was sizable, employing 5-10 different locals as drivers and guides, had a fleet of vehicles, and was "fully booked. We are doing good!"

I found many local companies attracted local people to work for them. One of Ari's guides, Hrefna, a local woman who grew up in town, told me she left a larger, Reykjavík-based glacier company to work for Ari's company because it was local.

Talking about her old job she said, "I felt like I work in a factory. When I went out on the glacier, it was all about taking as many many people as possible."

Hrefna was pleased with her new job with Ari's company. "They allow me to have a lot of free hand, and help them because they are pretty much starting with nothing and do not have very much experience from the industry. It's really nice and it's a bit more local and they want us guides to be part of the company, more than just employees." "Family" is how Hrefna described her new job.

One local man, Njörður, who periodically worked for Ari's company told me: "I almost lost this in the Collapse," gesturing around his home where we were meeting. "After, I worked here, here, here, and I fished, but I nearly lost it again. You know? And now, I work as a guide when I want to work, and I have not lost this house after two years in glacier cave guiding."

None of the companies "own" the ice caves—the caves are within glaciers inside Vatnajökull National Park's boundaries. During the time of my fieldwork, the industry was self-regulated, with no formal rules or standards for ice cave operators set down by the national park, the municipality, or the Icelandic government. Companies could operate at will—in any of the local glaciers' ice caves.

At the time, Breiðamerkurjökull had two accessible ice caves, and Fláajökull had a single ice cave that was accessible to tourists only after a rigorous and dangerous hike in the winter. A farmer living near Fláajökull, Kristófer, told me, "We are watching Fláajökull for a potential better ice cave to open. It will make the value of the area increase, and more people will go into the area. Right now, though, the guides have to be on their toes about where they take people."

Kristófer was referring to the dangerous hike through a rock fall zone to get to the one ice cave on the glacier, and his hope that better ones would open soon. Farms operating guesthouses near glaciers with ice caves would see a significant boost in guest numbers and purchased excursions.

While no company owned an ice cave, locals reported *feeling* a strong sense of ownership over the ice caves and local glaciers. Such feelings were evident in how locals talked about non-local companies and the state of tourism within ice caves today. The expressions of ownership were also suggestive of strengthening connections between people and ice across the region.

There was a firm distinction in local conversation about which companies offered ice cave tours—local companies or "Reykjavík" companies—and where each company sat within the social hierarchy of ice cave operators. Torfi, a twenty-something young man, guide, and owner of a local company, articulated what many other guides and local owners expressed to me: "The new owners of the Reykjavík companies are just old bankers from the crisis...They are just making money machines, not for the ice, I know they have lowered the prices by twenty percent and they are now hiring foreigners, their friends from England and New Zealand because those guides get paid less."

Torfi's comments require some unpacking. A perception locally fixates on the idea that ice cave tourism was started by local companies, and, as such, larger Reykjavík companies are seen as pushing in, unsafe, and ruining the experience for others. Local companies, Torfi's comments imply, are not just operating for the money. They are operating because they *want* to work with the region's glaciers—a notion I heard echoed by many others in the area. Hence, Torfi described the Reykjavík companies with the dismissive term "bankers," touching on palpable resentment simmering in Icelandic society towards the financial institutions that, through mismanagement, caused the financial collapse.

The addition of the 'non-local' companies into the ice cave industry has only really occurred over the last year, and many locals were incensed. Torfi furthered explained: "There were three companies with [a local company] in the beginning working together to arrange the ice caves, trying to balance out so the groups don't overlap and it does not get crowded in the cave. [Guide company] made a schedule, and for three years that worked. Until this year."

Dagfinnur described the comradery he felt at the end of his first season doing ice cave tours. "I am more confident. I learned from the other companies. They each learned from the company before to Einar. It is always hard being the new kid on the block. I made mistakes, but I learned now."

But with the addition of the Reykjavík companies, I witnessed the situation growing chaotic with mounting pressure.

Torfi noted that while his company used to offer other trips in the winter, such as northern light tours or glacier hikes, it was the ice caves that were selling and now his company only had the resources for that. His company was booked solid for the rest of the season and into 2016-2017: "We turn down ten people a day. We're totally booked... It is just getting crazier and crazier. Last week, there were over a hundred people in the ice cave at once."

A woman of prominence in the local community, Lovísa, described how much pressure she thought local companies were getting from Reykjavík companies. "I know that they are getting so much pressure, like, hundreds of people, like two hundred people had booked an ice cave trip TODAY. Even if there is no weather to allow safe access, like the weather yesterday, it was crazy to try, but the companies go anyway. It's so much money. The Reykjavík guys are like, hard on the local companies they booked with—you can't cancel, you can't cancel! But when there are so many people there in the ice cave, there is nothing to experience."

What Torfi and Lovísa refer to are practices from previous years, where Reykjavík tour companies would sell ice cave tours, but the local companies would physically operate the tours. However, in the last two years, the demand has grown so high that the Reykjavík companies are now operating the tours themselves, stretching the capacities of the ice caves to the limit and causing local conflict.

Ari's employee Hrefna summed up the situation colorfully: "If you look in the ice caves, it is total chaos, people are fighting. The bigger companies come in when they want instead of working together and just being professional. I feel like it is like this in everything in Iceland, the focus more on who has the biggest penis, focus on the competition, who has the biggest balls."

As I continued to visit the ice caves throughout the winter, it became increasingly evident that business was growing at a rate the ice caves could not handle. Groups were overlapping, with too many people pushed into small fragile spaces at one time, jostling for photographs, etching their names into the ice, leaving litter, and in a space with little air circulation, creating a distinct odor of overheating bodies. But each one of those people in those ice caves represented a great deal of money, of economy, to many people in the region.

Lovísa found herself repelled by the recent condition of the ice caves: "It's totally crazy. It totally ruins the experience of our ice cave. Because an ice cave has to be experienced in a small group of four or five people, but, when you are herded into it—herded in with a hundred people or more at a time, it is crazy. These are not our experiences we are selling."

Lovísa's sentiments were echoed by many others in the industry:

Elín observed, "I think it is a time bomb, right now. This winter is the largest one in our ice cave trips."

Thelma said, "I think it is not a good development. Too many companies, and the problem is that the old companies that had people who had been educated, who had taken courses in ice climbing and climbing and using the equipment, but many of the new companies they are just going in there and they don't have any knowledge about our glaciers."

Hreiðar said, "I say that even though it was amazing when this wave went off for the ice caves, when Einar starting doing this tour, and everyone put their money in and now has a big jeep, but it's just one ice cave, and if everyone wants it, I at least don't want to be in the middle of an ice cave with fifty people."

Ástþór laughed and explained, "I don't know if next winter there will be a cave, but if there is, I have booked seven tours for next year. I have now already spent that money!"

Elín thought too many people were in the ice caves, Thelma was worried about the level of environmental education occurring, Hreiðar had memories of before when the ice caves were calmer, and Ástþór was already both selling tours for next year *and* spending next year's money—and he had no idea if there will even *be* an ice cave next year.

The opinions were but a few of the many, many people who spoke to me about the local ice caves. Everyone in the area seemed to have opinions and wanted to share their thoughts.

What was revealing, though, is that stepping away from obvious issues and opinions concerning safety, lack of regulation, education, or predatory business practices (all important things, but not the primary focus here), what became apparent was how entangled local people were with ice caves. How much they cared about *their* ice caves.

Local people clearly perceived that ice caves were decidedly local—the ice caves belonged to the local companies, and by association, the people who live in the area. It was they who began ice cave tourism in the country, and it was they who must speak for the ice caves. Locals did not say local companies are bringing in too many people, or vandalizing the caves. The conflict was not over money—the local companies were fully booked and at capacity.

Locals held opinions about the best way to experience *their* ice caves, an experience that was repeatedly described to me as un-urban. Some people articulated that companies from

Reykjavík might be fine with crowding, but local companies—used to the region's isolation, fewer people, and no crowds—held different views of what was best for the area and the glaciers. In many ways, it was not that the nonlocal companies were bad, it was more that they were doing it wrong—and this impacted local companies who were doing it right. Because locals perceived ownership with the ice caves, they perceived themselves as positioned to determine the right or wrong way to interact with their caves.

IN LINE WITH MANY PLACES IN THE WORLD, Iceland experienced a significant rural to urban migration, notably centralizing two-thirds of the country's population around the capital area after WWII.[62] In the 1990s, a second wave of migration, especially of women, occurred, due largely to socio-economic changes including shifting rural fishing industries, access to secondary and tertiary educational centers outside the capital, and changing gender values.[63] Such trends continue today across rural Iceland, where young people grow up and head to Reykjavík in pursuit of diversified job and educational opportunities in the capital region. Most do not return to Hornafjörður.

Migration is not just limited to young people. In Höfn, many older people observed to me that as their children—who lived in the capital—started to have kids, they also chose to move into the city to help with the grandchildren.

Glaciers tourism—especially, ice caves—offered a different model. Bryndís, one of Ari's business partners who grew up on a local farm, explained why she became involved with the glacier industry: "I'm not going to spend my life waking up around six o'clock to go milk cows and go to sleep around midnight every day, always the same thing each day. And my parents they are doing this very well and they are getting old and I don't want to do that." The way she saw it, the choice was move to the city or do something else. She chose glaciers and starting a business. Now she travels and works on the business and wakes up when she wants.

Ástþór, who owned his company with his wife, told me how when he was younger he dreaded his two children getting educated because to him that meant they would eventually leave Hornafjörður. Now, he told me, his son was going to start driving the following year for Ástþór's company, and his daughter helped with the bookings. Ástþór was happy that he would be able to work in business with his children.

All seven of the larger local companies were family owned and operated—and they hired many local people. Árvök, a local woman who went abroad when she was younger for work and did not return to Hornafjörður until in her late forties, told me that she was finally able to come home to Höfn when tourism took off. Her parents were ill, and she worried about not being able to return without a job. But the glacier tourism provided her with the means to come home.

Þórhallur, a local man in his early twenties, had financed and built his first hotel the previous year. He saw the opportunity to stay on his family's farm in a new way with the tourism boom.

Stories like these abounded—each individual in details, but in general they were stories of people staying or returning to the area because of glacier-related tourism. Personally—it was also conspicuous: for the last nine years I have been spending substantial time in Höfn, and over the last two years there were noticeably more families, young children, and younger adults.

Still, the situation is quite complex. Many local people perceived some form of the idea that the creation of ice caves was linked to glacier melt, and subsequently, to climatic changes. As in, many people who spoke with me concluded that with increasing temperatures (brought on by climatic changes), glaciers would melt more, and a product of that melt process was more ice caves. And, returning to the opening of this chapter, a widely held belief was that more ice caves meant more money.

As the glaciers melt, they melt money.

Representative of such perceptions was Thelma, a local woman who explained: "Breiðamerkurjökull is hotter, and more ice is becoming more water, and the water is cutting under the glacier making the ice cave caverns inside. I've seen the water pouring down. This is just going to happen more and more."

There is some debate in glaciology concerning formation of specific features utilized in tourism as ice caves, and if indeed increasing temperatures will cause the features to be more or less common. Essentially, there are many variables that influence melt rates and the formation of englacial and subglacial channels (the most common "ice caves"). What recent research in Svalbard suggests is that with increasing temperatures, more meltwater is produced, which does indeed carve out more and larger features.

However, such features tend to disappear both as the glacier recedes and as ablation on the surface of the glacier increases with higher temperatures.[205, 206] Hence, potentially fewer ice caves. One researcher told me: "What we've seen is that with warming, the cave systems pretty dramatically collapse and are no longer accessible." This is in direct contrast to local perceptions, which understand higher melt rates to produce more ice caves.

Really, though, the details are not the point. Rather, what matters is that locals perceive ice cave creation in specific ways, and connect ice cave formation to larger forces including climate change.

This results in complex responses to the current state of

affairs—i.e. the financial benefits of ice caves. As Hulda said, "We are told it is not good for Iceland, but Breiðamerkurjökull's walking back I don't think it feels bad now." Paraphrasing Hulda: Icelanders are told glacier change is bad, but it doesn't *feel* bad yet.

Some locals perceived that they might be profiting in some way from climate change. For example, Finnborg echoed many other Icelanders when she said: "I have a feeling that Icelanders think, like my father, oh, this climate change isn't going to affect us that much. Glaciers, we go, we know that glaciers change. Sea level rise and warm temperatures, this is good for us, we might have warm vacation and money in the bank and more grass in the field and so on, because the possible changes for us on the small scale are not so big, so then we are not afraid about climate change. I feel guilty saying some of these thoughts."

Finnborg listed some of the perceived advantages of climate change in the area, including smaller glaciers, warmer temperatures, and more money in the bank through ice caves. But she also expressed guilt—after all, climate change is largely framed as a negative experience, so what happens when it doesn't *feel* like a negative experience?

Marín, a good friend of mine, expressed something in line with Finnborg. She said, "I had never given climate change, or the glaciers' departure, or the fishing development due to climate changes any much thought and I sometime read about it in the paper here, the Icelandic paper, but I don't read a lot from England or the BBC. I only follow the Icelandic news and sometimes there are reports about climate changes but they are very dramatic and sometimes they are so dramatic that I just turn it off, you know. Because that is not happening here yet, except in that the tourists come here to see that, you know, in the ice caves. The tourists are paying my brother [who is a guide] to show them climate change [in the ice caves] which I think is a little odd, but we can go on vacation."

Marín said she does not pay attention to dramatic news of climate change elsewhere, and knows that climate change is not happening in Iceland—not yet. Except the ice caves. Marín linked climate change and ice caves, and then talked about how her brother showed foreign tourists climate change via the ice caves. Marín felt a little odd about that, but was excited that her brother would be able to take them on a vacation soon. Marín talked extensively about her upcoming trip to Europe, and did not mention climate change again. When she eventually went on her trip with her brother and returned to Iceland, we sat together for two hours looking at her vacation photographs. She never mentioned climate change or glaciers.

Not everyone is packing for vacation; there are certainly additional costs to the increases in glacier-related tourism. With one grocery store servicing the entirety of Hornafjörður, many locals complained that shelves were often empty because of tourists. Local people employed in medical services noted they could not keep up with demand thousands of tourists brought to the region.

For example, Elín, a director of the medical facility, talked about the region's extreme strain: "This clinic is supposed to respond to the 2,000 inhabitants. And now they are talking about 4,000 people tourists staying here overnight, every single night. The high season. And the medical system is not built for that."

But, Elín hastened to add, it was not a problem easily solved: "It is not like we can add more doctors or nurses. First, we have to find houses for them, and all the houses are for the renting to tourists now."

A large portion of the region's medical services are supplied through a volunteer search and rescue (SAR) network. If an injury happens on the glacier, local people volunteer to help. This system worked when there was a low call volume, but with increasing tourism, volunteers were summoned more often, and the costs of such rescues or medical interventions fell upon the municipality and not the recipient of care. Locals were troubled by this problem.

For example, Lovísa observed: "People are getting tired. It's crazy, almost every day, sometimes like two times a day. Almost a full-time job for volunteers. And nobody is paying them for it because they are just volunteers. And at some point they will just burn out. Many of them are in the tourist business themselves, so tourist time is also a crazy time for all the accidents that are happening."

And increasingly, local people perceived their volunteer work not as helping their own neighbors, but helping foreigners who were not respectful of Iceland. Lovísa continued, "We [SAR] have not been called out much for Icelanders. It is actually getting to be a rare thing to help Icelanders in danger or need, so almost every call is for foreigners, like up on the glacier, usually it is some tourists on walking coming like across the glacier and not paying attention to safety."

Another guide, Klængur, echoing many people in the region, argued for a better system for tourists: "We need to find some way to control the traffic on the glaciers. Too many people are going there without guides, without any knowledge, and one of the problems in Icelandic tourists today are the accidents, like the accident in Reynisfjara two weeks ago… And the search and rescue are now too much, too busy. I think we should make the tourists more responsible for their own risk in Iceland."

Many problems were related to increased tourism in the region. However, few Icelanders told me they wished tourists would stay away. Rather, Icelanders were struggling to adapt to these new conditions, just as they are learning to adapt to transforming landscapes.

. N .
V ✦ A
· S ·

HOW ARE WE TO LIVE IN THIS WORLD? Glacier change is talked about as a largely negative experience for people and communities, and likely, it feels like that to many. I know that looking out the window at glaciers I've known all my life who are in recession feels devastating.

But, as seen in Iceland and likely in most near-glacier communities, ice loss is not a devastating single story of loss. Of destruction, of melt. Yes, that is all part of the story, but it is by no means the entirety of the story of people and ice across the world.

Part of the story of glaciers today is the short-term and long-term material advantages and disadvantages that shift each day dependent upon the needs of individual Icelanders. Three to four years ago, thousands and thousands of people from all over the world started to travel to Hornafjörður to experience ice caves—and see climate change. Alongside this influx, the region's post-Collapse economy started to thrive, people started new businesses, secured financial stability, and learned how to navigate increasing conflict and strained social services. And paradoxically, people also started to spend more time with the region's ice, increasing their local knowledge and experience with glaciers, and cultivating stronger senses of ownership—all at a time when the ice itself was rapidly dissolving away. Icelanders had to learn to make sense of their glacier prosperity amidst a global chorus articulating glacier loss as fundamentally negative.

If we are to understand how people live in a world *with* ice—this very moment—and what might happen as we lose our glaciers, it is essential to see the complexities, and see how people and societies transform with the positive and negative together. Iceland has been experiencing climatic shifts over the entire length of human habitation, and "climate change"—as in, global warming—for over a century. Any Icelander alive today has lived through a time of intense glacier change; glaciers have been changing their entire lives—and those experiences have been beyond positive or negative.

While near-universal evidence tells us that climatic changes are occurring, and that one of many impacts is the reduction of the world's glaciers, the processes underway do not determine the futures of any human or thing on this planet.[207] Glacier change implicates a host of complex, interwoven physical and social changes, all of which fluctuate at multitudes of scales and timeframes. A specific impact (of glacier change/of climate change) gains a value when it is plucked out of the entire system in which it moves—thereby disentangling it from the entire process.

Nothing is predetermined. Nothing is assigned a particular value.

When we got ready to leave, Snorri, Jack, and I stood at the entrance to Breiðamerkurjökull's Crystal Ice Cave. We could see back into the blue, and we could see out into the world, where it was billowing snow.

Snorri's jeep was parked just a short distance away, yet it looked to be in another world.

"Thank you so much for getting me out here," Jack told Snorri.

Snorri nodded.

"Back home," Jack told us, "We have these dark-colored butterflies. I think they're swallowtails? They're black, but they have the most iridescent blue edges on their wings. Just magical. I feel like I've just been inside butterfly wings."

Snorri grinned.

"I like that!" I said.

"But," Snorri added, "don't butterflies die when you touch them?"

Movement caught our eyes, and, turning away from the ice cave, we saw bright headlights bouncing down the sandur, more tourists heading to the blue light.

CHAPTER EIGHT

≈ ≈

WHERE DO WE GO FROM HERE?

As I write this book, the headlines of several international newspapers echoed a story reported in *The Guardian*: "Receding glacier causes immense Canadian river to vanish in four days."[208]

For hundreds of years in Canada's St. Elias Mountains in the Yukon Territory, meltwater from the Kaskawulsh Glacier flowed for fifteen meandering miles along the Slims River before merging into the Kluane and then Yukon Rivers, and flowed for another two thousand miles across the crest of Alaska all the way to the Bering Sea. But in less than four days in 2016, the rate of the Kaskawulsh's recession accelerated and the glacier redirected its meltwater south, down the Alsek river for 240 miles straight into the Gulf of Alaska.

The Slims River was over 500 feet wide in some places, and now it's gone. Waterfront communities, including Burwash

Landing and Destruction Bay, look out now on exposed riverbed sediments and miniature hoodoos. In news videos showing the vanished river, local people look around and shake their heads in disbelief. In the preserved oral history of this region, rivers have never disappeared entirely. For that matter, neither have glaciers. What do local people do?

This is a book about Iceland, but it is also a book about what it means to live in the modern world during a time of immense change. And we are losing our ice.

Hornafjörður native Dagfinnur told me: "From May to today [October 2015], there are areas that are just gone. We were looking at them five months ago. And now they are just completely gone. We were walking and then there are glacier areas that are just gone. Completely gone. It is really hard to predict what the glacier will be tomorrow."

In the first chapter of this book, I said that I would not present just another explanation of the ice we are losing, but rather that I would show what we may yet find *with* ice—all the complex ways people and ice connect, overlap, influence each other. The problem, of course, is that while glaciers are part of vast evolving communities of ice systems and landscapes and climate and people and cultures and histories and futures—and while glaciers structure river systems and lichen colonies, nationalities and identities, senses of aliveness and living, lifeways and economies, floods and fish—and while glaciers are more than the summation of their parts and processes, more than climate change yardsticks and thermometers, more than melt and "just ice"—and while glaciers have rich social and cultural context and variability, are complex, contested, controversial, and are profoundly entangled with human and more than human lifeways—we are still losing our ice.

So then, where do we go from here? What do we hear when we listen to glaciers? What lessons can we take away, and how might the many people worldwide living without glaciers in their backyards learn from ice as well?

. N .
V ⊕ A
. S .

ALLOW ME A WEE JOKE: this book is just the tip of the iceberg.

What I have explored in these pages are the rich and profound connections between people and ice in Hornafjörður, connections that shape how people understand and relate to their surrounding environments. I chose to focus on this region because, following Iceland's preeminent glaciologist Helgi Björnsson, "Nowhere in Iceland has the proximity and relationship between man and glacier and its rivers been so intimate and difficult than along the coastline south of Vatnajökull."[3] Nowhere have I felt such a draw to sit day after day in the laps of glaciers, gazing in heartfelt wonder at such enormous icy creatures.

In Hornafjörður, people and ice have been interconnected since Settlement, a connection still strong, diverse, and transforming today—over a thousand years later. Premised on a belief that knowledge is produced and not revealed, and that the production of knowledge is often shaped through narrative, I traveled for years across the region, pursuing a variety of glaciers stories that were produced, circulated, and authorized by local people. I sought to understand both how glaciers entangled in the everyday lives, practices, and knowledges of local people, and how ongoing climatic changes might transform those relationships. What I found and explored in this book shows both immense glacier diversity and immense human diversity.

I found that through practices including formal and citizen glacier science (the production of glaciological narratives), Icelanders over the last several hundred years have woven glaciers into their cultural fabric. That in present times, glaciers command a robust position in Icelandic culture, suggesting that what happens to ice happens to an entire people.

I found that historical and cultural legacies of glaciers influence the ways in which people interact and think about local glaciers today. Examining how these glaciers—or forests, oceans, rivers, and other changing environmental phenomenon—refract cultural perceptions of change reveals both the nature of human-environment interactions and how such interactions might transform under the increasing stress of global environmental change.

In direct contrast to dominant framings of glaciers as remote and on the periphery of human society, I found that glaciers were powerful participants within the community of Hornafjörður. Glaciers possess immense power that materialized through glacier floods, destroyed homes and livelihoods, and pushed people to make decisions: go, or stay. The narratives of two local farms, Heinar and Haukafell, vividly demonstrated how local glaciers shaped people's senses of movement, space, and scale. And, again in contrast to dominant framings of melting glaciers as widely lamented, I found that context is crucial for local understandings of glacier change. Many Icelanders expressed relief that the ice was disappearing, believing that as the ice recedes, so too does the threat of sudden destruction of home and family.

The implications of these findings are critical for understanding the diverse ways people experience environmental change. Demonstrating glacier power reveals how different environmental phenomena enact immense power over their surroundings (including people), which circuitously influences how people react and respond to change. No person is an island, no glacier is a single unit. While glaciers are receding at a global scale, glacier power differs from place to place and the experiences of glacier change are inherently local.

I also found that people consciously or unconsciously shield themselves from difficult realities. That while glaciers across the southeastern coast were diminishing, such change produced different stories with diverse meanings, codes, and assessments of short and long-term advantages and disadvantages. This was due in part to the generations of Icelanders who survived in tandem with the ice in this region, guided by the glacier narratives they told each other over and over again. These cultural stories of glaciers shape how Icelanders see modern glacier change—just as people worldwide learned the rhythms of their own backyards, rivers, and forests. Just as you rely on your own local stories to understand what is changing, and how it is changing. Just like we all do.

Today, people worldwide rely *first* on our powerful cultural stories of change as lenses through which to understand and articulate the meaning of the almost unimaginable environmental change happening in every backyard. And if human society—you, me, all of us—realistically wants to move forward with effective strategies to meet climate change, and we realistically want diverse peoples worldwide to understand and participate, then it is essential to begin paying attention to the cultural stories people tell about their backyards. It is these stories that often determine what people can and cannot see, and how people fit themselves into larger processes at play. These stories might be about negative impacts, or positive impacts, or impacts somewhere in between, but critically, these are stories of transforming environments, and we need to listen.

In Iceland, glaciers possess a plasticity to verify multiple conflicting narratives all at once. Consideration of this dynamic tells us that context is crucial, and that there are simply no single stories of ice—nor are there single stories of anything else. Understanding this tells us that just as people's relations with the environment are multi-storied and complex, so too then must be the approaches and strategies for responding to environmental change. And critically, such approaches and strategies must be grounded within the human story of specific landscapes, contextualized within specific places and times.

I found that we can have the very best data, statistics, and models chronicling glacier change, but if that information is not grounded within the human stories of place, then that information is largely powerless. If people do not see themselves in the story, then they are not a part of the story.

I found complex perspectives of glaciers alive. While such perceptions strengthened relations between people and their environments—people cared about ice, and ice cared about people—they also pushed against dominant ideas of what glaciers are and can be. I found people often possessed multiple knowledges of ice, and that such knowledges did not need to be, nor should they be forced to be, compatible.

I found a complex balance between loss and gain as glaciers recede in Hornafjörður. Over the short-term, glacier loss was perceived by some in Iceland as beneficial. Glacier recession reduced threats to many people and lifeways, but most importantly, glacier loss was serving as a financial boon to some local people through glacier-related tourism, especially in relation to ice caves. Through glacier tourism, in just the last couple of years, many Icelanders have regained post-Collapse financial security.

These glacier change realities demonstrate just how influential local political and economic circumstances are in shaping how people experience environmental change.

As such, it is critical to represent, understand, and acknowledge the harrowing complexity and immense value range of lived people-ice experiences and interrelations in diverse places such as Hornafjörður. Not doing so treats glacier change as a geographically undifferentiated phenomenon devoid of localized, grounded, and relational experiences of change.

The implications of the findings laid out here are intricate. First, it is essential to understand that people will comprehend issues such as glacier change against both a historical and present-day backdrop—and by overlooking such socio-political context (and place, and time, and culture), it becomes near impossible to understand in any meaningful way the complexities of any given issue. People who seek to reduce and simplify such issues are only trying to control the narrative—frankly, issues in our world today are not simple. They're excruciatingly complex, and great patience and care needs to be utilized as we seek to understand them.

Second, applying wholesale valuations for any particular issue is limiting. That some Icelanders find short-term benefits right now does not negate some of the other very real challenges of continued glacier change on the southeastern coast. Reducing glacier change to a particular value or good or bad then overlooks the transient experience of all change. Focusing on just a handful of specific outcomes or experiences will likely fragment community buy-in and support. Rather, recognition of the ongoing ephemeral nature and range of all environmental changes—and concerted efforts to communicate all characteristics and experiences of change—will likely engender broad community support for future strategies.

SO AGAIN, WHERE DO WE GO FROM HERE? The Kaskawulsh Glacier is disappearing. So too is Skálafellsjökull. The eastern side of Hoffellsjökull was once an entirely separate glacier—but local dynamics withered that trunk of ice down to a nondescript stump. Fláajökulll is thinning and receding, sculpting the mountain Jökulfell in the process. Heinabergsjökull thins like an undernourished pancake, but maintains a steady terminus face. And Breiðamerkujökull—one side of the glacier's snout face is receding back while the other half visibly distorts.

I've been to so many of Earth's glaciers, and I've never seen the same glacier. Each glacier is different, changing, fascinating, worthy, and important to our world. But what is striking in Iceland is the immense variety of stories between people and ice, stories predicated upon the individuality of the glacier, person, community, and region.

Allow me to belabor this argument for one moment, as it is essential to understand.

The chapters in this book examining diverse relations between people and ice show that no single story of glaciers exists. Glacier stories in Iceland are multitudinous and diverse and complex and contradictory and place-based and simultaneously rooted in time *and* timeless. I hope you come away realizing that people think all sorts of things about ice. Each glacier in each chapter responds to climate change in diverse ways. Through these diverse stories, glaciers are continuously imagined and reimagined, negotiated, transformed, assembled and reassembled, powerful.

Paying attention to the countless stories people and ice have produced together in Iceland reveals how people make sense of the changing environment. And this dynamic is more important than ever in today's age of the Anthropocene, where unprecedented changes to planetary systems are reshaping the very core of human existence on the planet.

Glacier stories open windows into the theatre of living, offer indications not only into what is happening on stage, but also behind the scenes. As researcher Anna Kaijser observed in Bolivia, "[w]hile the melting of glaciers is a physical process with strong material impacts, it gains meaning for humans through narration."[124]

Words about glaciers make glacier stories, and glacier stories make realities and viable futures. Stories are composed of entangled frames, associations, connotations, memories, values, beliefs—and people's thoughts, dreams, hopes, and fears. Every society at every time in every place tells stories, *narrates.* As American-Canadian writer Thomas King has observed, "the truth about stories is that that's all we are."[209]

People tell glacier stories about themselves, each other, the past and the future; right now, people tell glacier stories that are true and stories that are not, people tell glacier stories to pass the time and to accomplish tasks, to create direction and process emotion, to make order in the world and to create chaos. People create glacier stories to make sense of the world—the world of the human and more than human, the entangled world of people and nature and you and me.

In Iceland, glaciers breathe in, glaciers breathe out. In winter, glaciers breathe in snow, particulates, weather, atmosphere, skis, bodies, airplanes; in summer, glaciers breathe out water, sediments, mass, chemicals, algae, hiking gear, husbands. Breathe in, out, in, out. Except, now, they say, glaciers can't seem to catch their breath.

Stories of glaciers circulate in Iceland. Stories of glaciers and identities, of glaciers shrinking and growing, of glaciers alive and laughing, of glaciers destroying and creating, of glaciers saving people's lives. Stories abound. And each story tells us a little more about the storyteller, the listeners, the time, the place.

I HAVE WORKED TO HELP PEOPLE understand people and ice, focusing on local perceptions, values, beliefs—of how and why people and ice interrelate, and of how people make sense of melting ice. But this book only explains one place—Iceland—and one time—the present—and does not reveal the stories of people and ice from other places in the world.

Place matters for people and ice. Examining Iceland against the backdrop of the larger world confirms there are unique and diverse relations amongst people and ice in other places—that there is a global geography of glacier change.

We just don't know what that is yet.

As such, this book is not an end, but an invitation for more, an invitation to examine what relations exist between people and glaciers everywhere. Just as glaciers are diverse, so too are the human communities living alongside ice. Everywhere there are glaciers on this planet, there are people, and the two have been interacting in largely untold ways since the beginning of human history on Earth.

Given the number of glaciers in the world and the populations living near them, it is surprisingly difficult to assess how common or unique the types of relations between people and ice that exist in Iceland are in comparison to the rest of the world. Beyond a handful of studies, little is known at a fine resolution, and broad trends are just that—broad, with few details that can facilitate comparing experiences from one glacier community to the next.

I want to know how stories of Iceland compare to other glaciated countries in Europe, or in the Western world. What about people and glaciers in other places? While a handful of glaciers in Kyrgyzstan, New Zealand, coastal Alaska and British Columbia have been studied,[18, 148, 210] what about the glaciers and people in the Qilian Shan, Kunlun, or Altai mountains? How do people interact with the Buluus glacier, which is formed by frozen groundwater 100 kilometers outside of Yakutsk, the capital of the Sakha Republic (Yakutia)? Colloquial reports tell of emergent mammoth bones, of tourists being guided by traditional reindeer herders, of local debates over whether Buluus can even be classified as a glacier.

What about the estimated 140 glaciers in the polar Urals, such as the Institut Geografii Akademii Nauk (IGAN) glacier outside the coal-mining city of Vorkuta?[211] In what ways do local people (if at all) identify with the ice—and how might they compare with other people and places?

What about the Irik, Kokurtly, Bolshoi Azau glaciers of Mount Elbrus in the Northern Caucasus? Those glaciers are locally infamous for generating devastating ice avalanches, yet little else is known about how people relate with the ice beyond the thousands of people climbing the summit of the mountain each year. And while the media focuses on the glaciers of Mount Kilimanjaro in Tanzania, what about the eleven glaciers on Mount Kenya in Kenya and the thousands of people living in their shadows? The Speke, Elena, Stanley, and Baker glaciers in the Ruwenzori Mountains are largely absent from our stories of ice—from the public imaginary. So too the small Komando and Sannomado glaciers on Mount Tsurugi in Japan, the Heard Island glaciers, and Colorado's fourteen small glaciers. With the ongoing conflict in Kurdistan, do people relate any differently with southern Turkey and Iran's glaciers?

And what is "different"? If no baseline exists for what people have done in the past or at present with glaciers, how can we even begin to understand how people change as their ice changes? In Iceland, I found that glaciers were part of Icelanders' identities, and I pointed to the role of ice on the nation's flag. How does this compare to, for example, the Armenian coat of arms, which features the glacierized peak of Mt. Yervan at its center? Or the Greenlandic flag, which depicts the island's enormous ice cap? Or the US state of Colorado's flag, which features a white stripe to represent snow and ice? Even the city of La Paz, Bolivia, which sits in the shadow of glacierized Mt. Illimani, puts the mountain's glacial image on the city's official shield, university's seal, and the label of the local beer. Do all these examples suggest glaciers might be part of these diverse community's senses of identity?

In Greenland, glaciers melt so rapidly locals must revise maps, charts, and social memory to include recently uncovered mountains, islands, valleys.[212] Social knowledge—an important aspect of cultural identity—is shifting there. What impact does this have in Greenlandic society—and how does it compare to shifting knowledges in Iceland or elsewhere? There are over 450 glaciers in populated Kamchatka and some research regarding their physical conditions. But historical reports of these glaciers are at best over a hundred years old—and Indigenous Peoples such as the Koryaks and Evens have inhabited this region for centuries. Where are their stories? How has the ice—if at all—shaped their culture?

Given the prominent place of glaciers in the human imagination and cultural consciousness of climate change today, the gap in knowledge is concerning. When the findings of this book are juxtaposed against the glacierized communities suggested in this section, what distinctions amongst people and ice might come to the fore? How have different people in different places navigated change? What does a geography of glacier change tell us? How exceptional is Iceland?

Anthropologist Sarah Strauss reminds us that culture is the "*primary adaptation* of the human species,"[207] simultaneously connective and malleable to ongoing dynamic environmental and social changes. Glaciers are vital elements of Icelandic culture and the Icelandic landscape; adaptation to glacier change is negotiated there in both abstract and material ways. Some Icelanders have adapted to glacier change by constructing their homes atop hills in Mýrar; others have overlooked *unprecedented* glacier change by pointing to local histories of oscillating ice. Guides adapt to rapidly receding glaciers—and growing proglacial lakes—by using zodiacs to transport clients to the glacier's edge. Adding a "boat trip," they can justifiably increase a trip price. Glaciers change, and the people living near them change too. Embedded in Icelandic culture is a naturalized sense of adaptation *with* the ice. Glaciers change, and so too do people.

What about all the other places in the world with glaciers? How do they—will they—transform? Adapt? Respond? When Icelanders talk about surviving glaciers, they talk about sustaining, adapting, transforming, resilience. Working *with* change. The cultural emphasis on survival authorizes and legitimizes *the individual shape* of adaptation and response to change—the ability of individuals to make their own stories of glaciers and transformation.

Many people today use glaciers' vulnerability to awaken communities to the dangers of climate change. I suggest something different.

Every single place on this planet, every single person, every single everything, is transforming today. Transformation unites us all. How we respond to that transformation defines who we are and what our future can be.

What if we argued that part of living in this modern world means living in a world with ice—ice that is not going or gone, but that enrichens our shared world and is worth fighting for? What if John Muir was right all along when he wrote, "but glaciers, dear friend—ice is only another form of terrestrial love."[213]

Seeing glacier vulnerability means not only seeing a glacier flood coming, but also seeing one's self within the causal loop of the transforming flood. What if we motivated change not through fear and loss, but through beauty and possibility? Through mutual transformation? What if we highlighted glaciers to show that the world is exceedingly more complex, and beautiful, and imaginable, than we've ever previously fathomed?

ENDNOTES

≈ ≈

CHAPTER ONE

1 Here are the full quotations: "[I]ce plays a critical and major role in setting the temperate of Earth's atmosphere and oceans, governing major weather patterns, regulating sea level, and dramatically impacting agriculture, transportation, commerce, and even geopolitics... If we do not act now, as individuals, as communities, as businesses, and as nations to slow and gradually halt the current meltdown, we risk destroying the very global systems that have enabled us to thrive and prosper" (Gore, 2010). "By one estimate, the 46,000 glaciers of the Third Pole region help sustain 1.5 billion people in 10 countries... these glaciers are receding at an ever-quickening pace, producing a rise in levels of rivers and lakes in the short term and threatening Asia's water supply in the long run" (Wong, 2015). "There are cultural impacts of glacier retreat. Many human societies have strong attachments to glaciers...These features have strong symbolic significance, and people identify with them... If a rather extreme parallel may be drawn, many people experienced deep distress over the attacks on the World Trade Center in New York not merely because thousands of people were killed, not merely because valuable property was destroyed, but also because of the symbolic importance of the buildings themselves..." (Orlove, 2008).

CHAPTER TWO

1 The Hoffellsjökull/Svínafellsjökull glacier complex is two distinct glaciers: Hoffellsjökull and Svínafellsjökull. For brevity's sake, unless noted, I will refer primarily to Hoffellsjökull/Svínafellsjökull as just Hoffellsjökull. To add additional complexity, the glacier name Svínafellsjökull is the same as the glacier Svínafellsjökull next to Skaftafellsjökull in Öræfi. Many people visit the headquarters of the Vatnajökull National Park in Skaftafell and then visit the glacier Svínafellsjökull. As such, many participants interviewed for this research refer to Svínafellsjökull, meaning the glacier in Öræfi. To mark the difference, I will typically reference Svínafellsjökull as Svínafellsjökull in Öræfi, or Svínafellsjökull and Hoffellsjökull.

2 Lambatungajökull is hidden and hard to access—so much so that it rarely was mentioned by local people and I visited the glacier only once during my fieldwork.

3 Viðborðsjökull was inaccessible during field research due to a landowner's choice to withhold permissions to cross his land.

4 Of note, all the Icelanders in this book whom I quote or describe are anonymized except where the person is a public figure such as the president of Iceland. Additionally, Icelandic naming customs are patronymic or matronymic. A child's surname is determined by the first or middle name of the father or mother, as opposed to Western naming practices that traditionally reflect family lineage. Icelandic naming practices also have a strong gender element: male names combine the name of the mother or father with 'son;' female names add 'dottir.' For example, Sigrún Sveinbjörnsdóttir's son's surname could be Sigrúnarson (literally, Sigrún's son) and her daughter's surname could be Sigrúnardottir (Sigrún's daughter). This makes it easy to know the gender of a person before meeting. However, my surname is Jackson, but I am a woman. Most Icelanders showed humorous confusion upon introductions. Throughout this book (as culturally appropriate) I refer to Icelandic people by their first name, except where clarification or exactness to a specific person is necessary.

5 See media coverage of the storm at: https://icelandmonitor.mbl.is/news/nature_and_travel/2015/12/08/iceland_sees_double_hurricane_force_winds

6 You can follow my colleague's images at @ThorriPhotoFilm

7 Eight Regional Glacier Groups include: the Vatnajökull group, the Mýrdalsjökull group, the Hofsjökull group, the Langjökull group, the glaciers of Vestfirðir (Vestfjarðajöklar), the glaciers of Norðurland (Norðurlandsjöklar), and the glaciers of Austfirðir (Austfjarðajöklar).

8 Other thermal glacier classifications include polar, sub-polar, cold-based, warm-based, and polythermal. Geographic classifications include tropical, polar, maritime, etc.

9 Approximately 24,500 years ago.

10 The Sagas of Icelanders (also called the Icelandic Family Sagas) are medieval prose works about Icelandic families during ca AD 870-1050. Comprised of approximately one hundred sagas and shorter stories (the Family Sagas number about forty), the sagas were written and compiled in Iceland during the thirteenth and fourteenth centuries. Much scholarship has been devoted to the Sagas and topics such as authorship, ethnography, literature, and oral traditions. Of relevance here, however, is the consensus among saga scholars that these texts contain vast social and historical information on early Icelandic culture. As Ogilvie and Palsson (2003) argue, the role of climate and weather in the sagas demonstrates the cultural significance and relationship amongst Icelanders and the environment. Geographer Mike Hulme (2009), noting that the landscape of the island was so closely impressed on early Icelandic culture that prying apart expressions of nature and culture is impossible, argued that the sagas "reflect the way in which Icelanders thought and talked about the climate and their relationship to it." Saga scholarship suggests depictions of the climate were used metaphorically to represent human emotions and events. Interestingly, local glaciers were mentioned and named in the sagas. For example, Balljökull was first mentioned in Bárðar saga Snæfellsáss, Grettis saga, and *Ármanns saga*; Eyjafjallajökull in Njáls saga; Myrdalsjökull in Lögmannsannáll; Snæfellsjökull in Landnámabók, Eiríks saga, Bárðar saga Snæfellsáss, and Víglundar saga; and Sólheimajökull in Konungsannáll and Sturlunga: *Árna saga byskups*. These named glaciers were not just passive features on the landscape, rather, they had integral roles in medieval Icelandic culture that affected saga characters and bespoke Icelandic worldviews and value systems.

11 The 1783 eruption at Lakagígar is also known as the Skaftá Fires due to the lava that flooded down the Skaftá River.

12 Also called Litlahérað, a pre-eruption name.

13 It covers over 500 square miles.

14 "Já" is Icelandic for agreement. It is also a conversation filler, a way of encouraging the other speaker to continue.

15 Women historically, in comparison to Europe, have had high rates of participation in the labor market. In 2011, 77% of women engaged in the labor market, of which 64% of women worked full time. In comparison, during the same time, 83.7% of men worked.

16 While Iceland is not a member of the European Union, since 1994 it has been a member of the European Economic Area (EEA). Iceland applied to join the EU in 2009, however, in 2013, Iceland suspended its application (formal suspension occurred in 2015). Politically, the issue is quite contentious in Icelandic society. The primary unresolved issue centers on the country's fisheries, and how joining the EU would potentially alter regulations and trade agreements. For now, Iceland is a member of European Economic Area, which excludes the fisheries. If Iceland joined the EU, it would be the smallest country in terms of population, but one of the more well-off. The International Monetary Fund ranks Iceland in 2016 7th behind Germany and Austria in per capita GDP. In 2014, the country reported rates of employment for people aged 20-64 at 83.5%, the highest in Europe (Statistics Iceland 2016). The country's economy as a whole centers on tourism, fish, energy production, and the financial sector, and operates on a Scandinavian social-market model predicated upon capitalism, free market principles, and a broad welfare system.

CHAPTER THREE

1 The ice caves are called the Crystal Ice Cave and the Waterfall Ice Cave. I describe these ice caves throughout the book but I do not provide exact locations for any caves. Ice caves are exceptionally dangerous, and should not be accessed without certified guides or specialized knowledge. Safe access regularly requires modified vehicles and equipment, including crampons, full body harnesses, safety gear, and ice tools, and given the recent tourism increase in Iceland, many tourists are trying to access ice caves without purchasing guides, exposing themselves to extreme hazards.

2 Svavar is a public figure, so I use his real name.

3 Breiðabunga is 4,986 feet.

4 Also spelled Nýju Núpar. Records show that Nýjunúpar projected through Hoffellsjökull's surface around 1910 or so as the glacier body gradually thinned.

5 Vigdís served as president of Iceland for 16 years, from 1980-1996.

6 Perhaps the best illustration of Sigurður Þórarinsson's prominence comes from the Geological Society of America's written memorial to him upon his death in 1983. They wrote: "Sigurdur Thorarinsson [sic] was in many ways the personification of Icelandic geology, and in his day, publically, by far the best-known earth scientist in Iceland" (Steinþórsson, S., *Memorial to Sigurdur Thorarinsson* 1912-1983. Geological Society of America: 1983).

7 Saltpetre, also known as saltpeter or salpeter, is a naturally occurring alkali metal nitrate (most often sodium or potassium nitrate). It is a white mineral that crystalizes on cave walls (accumulating in bat guano, for example), thus, during this time period, it is understandable why the early Danish thought glaciers might be made from the same white substance.

8 *Reise* translates to "travel" in English, and is referred to as *ferðabók* in Icelandic, meaning travel book.

9 Forty years after Pálsson's death, nearly a century after he wrote his manuscript, two chapters were published in the 1880s in Danish. In 1945, the manuscript was published for the first time in Icelandic in Iceland, and in English in 2003. The publication dates, and the language of publication, are important. The majority of Icelandic speakers reside in Iceland, with small pockets of speakers in Europe and Canada. As such, publishing Pálsson's work in Icelandic in 1945, while a feat for Icelanders, still sentenced his work to continued obscurity. As Michael Gordin (2015) notes, after the 1920s, the multiple languages of science—German, English, French, and Latin—condensed into an American-centric English. Works published in Danish, let alone Icelandic, were rarely registered in the scientific community.

10 Icelanders did not produce saltfish until the 18th century due to lack of salt. Once they acquired salt from Mediterranean countries, saltfish exploded as a primary export. Between 1920-1935, world production of saltfish peaked. For Iceland, Spain was the primary market for saltfish. However, when the Spanish Civil War broke out in 1936, the

trade with Iceland came to a halt. Exported fish was then primarily chilled with ice and exported to Britain.

11 The recession of the glacier lagoon in front of Hoffellsjökull is a recent development. The glacier reached its maximum extent in the 1890s, but did not begin pulling away from its eastern moraine until the 1940s.

12 Starting around 1897, fishermen took advantage of the single merchant store in the county located in Höfn to acquire goods and trade. In 1908, the first motorized vessels sailed into the Höfn harbor for the winter fishing season, and from approximately 1920 through 1948, Höfn acted as the fulcrum for motorized winter (February-April) fishing in the eastern fjords. People came from all over the island to live in Höfn and work in the fisheries. Due to poor fish processing facilities and the difficulty larger modern motorized fishing vessels encountered when accessing the village's harbor, the fishing industry moved away from Höfn from approximately 1950-1970s. The population of Höfn plummeted. However, the industry gradually rebounded after the 1970s, due in part to the construction of better facilities, more consistent dredging of the harbor, and notable proximity to Iceland's richest lobster and herring fishing grounds.

13 Women are not readily visible in the narrated history of glaciological science in Iceland. The historical documents I reviewed alongside the numerous stories I heard were notable for their absence of women. When discussing this with another researcher in Iceland, she observed: "It is my belief that the invisibility of women is not due to their scant knowledge of environmental changes. During my fieldwork I spoke to elderly women who lived near glaciers, and they seemed equally interested and knowledgeable about the changes in their environment as the men. It is more likely that the invisibility of women is due to the status of women in the mid-twentieth century [in Iceland]. Women farmers were housekeepers, mainly in charge of tasks in and near the home. The men however, traveled further from the home, such as during the annual gathering of the sheep in the mountains. Men also acted as the spokesperson for the household . . . I have discussed the gender issue with some of the experts that I interviewed. Most of them agree with my belief that environmental changes were observed by women as well as men, and discussed within the family.

However, women are not visible in the material because they did not act as spokespeople for what may have been (at least to some extent) the common ideas of the household. Moreover, it was only men who performed the glacier measurements for the Iceland Glaciological Society. . . . Thus, I believe that although women are not visible in the material, their knowledge is still present in it to some extent" (personal communication 9/29/2015).

14 Website: http://spordakost.jorfi.is/?lang=eng. Today, people all over the world can access the glacier monitoring website, click on any of the glaciers, and be presented with data from 1930 to the present day detailing the glacier's movements, historical images, present-day information, and the names and dates of individuals responsible for monitoring particular glaciers.

15 For an excellent overview of Icelandic glacier names, see O. Sigurðsson & Williams, 2008.

16 Oddur Sigurðsson is a public figure, and as such, I use his real name.

17 As these scientists have published their work in public forums, I use their real names.

18 I focus in this chapter on the connections between Icelanders and glaciers through art. However, there are many other connections, most notably, in literature. Many scholars have argued that Icelanders connect with the environment and with glaciers sustained upon centuries of narratives of people and ice interacting and transforming in concert. Such narratives have been preserved, developed, and proliferated in oral and written form. As early as the 13th and 14th centuries, Icelanders were writing about local glaciers in the Icelandic Family Sagas (Íslendingasögur). Folktales, ballads, and fables related to or involving glaciers have materialized across the island and are woven throughout modern Icelandic culture. Today, glaciers are found in Icelandic poetry and novels, in fiction (e.g., Sjón's 2008 *The Blue Fox*, and Halldór Laxness's 1968 *Under the Glacier*), prose (e.g., the *Útkall* series by Óttar Sveinsson, who has written one book each year in the series since 1994, many of which involve glaciers), myths (e.g., Runa Bergmann's 2014 *The Magic of Snæfellsjökull*), short stories (e.g., A.S Byatt's 2003 *The Stone Woman*), sagas (e.g., Icelandic Family Sagas), and plays (e.g., local plays held on Snæfellsnes Peninsula at

"The Freezer"). Icelanders, as has been noted widely in the media, are "prolific readers and writers." The Icelandic Literature Centre reports that one out of every ten Icelanders publishes a book—and many of those books explore Icelandic nature and glaciers to some degree. The best example is from Iceland's (to date) only recipient of the Nobel Prize in Literature, Halldór Laxness, whose quirky 1968 novel *Under the Glacier* depicts the story of a young man sent to a remote parish at the foot of the glacier Snæfellsjökull to report on the activities of the unconventional parish priest Jon Primus. Snæfellsjökull in this story looms over the village and the locals' activities, with each page conveying "the godly power of the glacier throughout, like a continuous flow of background music. . . . " Susan Sontag wrote, noting the dialectic between Jules Verne's use of the glacier Snæfellsjökull and Laxness's sly response in his own novel, that in "Laxness's story, a sojourn near Snaefells [Snæfellsjökull] does not call for the derring-do of a descent, a penetration, since, as Icelanders who inhabit the region know, the glacier itself is the center of the universe."(Sontag, 2005). Laxness uses the glacier in his novel to suggest organized religion does not have a monopoly on divinity and that sanctity might be just as likely located within surrounding environments. As the priest in *Under the Glacier* observes, "Words are misleading. I am always trying to forget words. That is why I contemplate the lilies of the field, but in particular the glacier. If one looks at the glacier for long enough, words cease to have any meaning on God's earth."

19 President of Iceland Ólafur Ragnar Grímsson served from 1996-2016.

CHAPTER FOUR

1 In Iceland, milder conditions previous to the LIA might have, in McKinzey et al.'s (2005) articulation, "induced a misleading sense of security for traditional farming practices, which meant that grazing regimes previously used for generations may have been poorly suited to LIA climatic extremes." Iceland experienced widespread farm abandonment during this time—and farms were not resettled until after the late nineteenth century, when glaciers began to recede and the climate returned to more mild conditions.

2 During this time, Heinabergsjökull was merged on its southern terminus margin with Skálafellsjökull.

The 1362 eruption of Öræfajökull and subsequent jökulhlaups flooded and decimated the region of Öræfi. The tephra from the eruption coated the south coast, and scientists use it as an easily identifiable layer in soil profiles.

3 Fláajökull is not the only glacier in the area to roll across farms. Most notably, Breiðamerkurjökull destroyed the farm Fell in 1869, an event still discussed locally. Other farms, such as Breiðá and Fjall, were also destroyed by glacier expansion. Reports suggest the Breiðá farm (from which the glacier Breiðamerkurjökull derives its name) was abandoned in 1698 and was entirely covered by glacial ice by 1712.

4 Leifur was born in 1917 and was twenty-one at the time, and his job every morning was to bring water to the kitchens by bucket. He remembered that the older men at the worksite called him "waterboy," which he didn't mind, because his wife was eighteen and working as a chef in the kitchen.

5 Between 1920-1940, Fláajökull was noticeably receding, and local people grew excessively worried. They thought at the time that the more the glacier melted, the larger and more catastrophic the jökulhlaups would be. The German government in 1930 sent a Zepplin flying around the countryside to both celebrate 1,000 years of Alþingi government, and to take pictures of the landscape.

6 In 1946, the people of Mýrar had to build brand new dikes, and again in 1958 and 1960. Smári told me as we walked his land that in the decades after 1952, he'd go walking and find all the things people used to put in the dikes—wagons, Model-Ts, carts, rock bins—washed up on the coast. People were willing, he said, to do anything to get rid of the glacier threat.

7 Sociologist John Law (2004) defines assemblage as "a process of bundling, of assembling, or better of recursive self-assembling in which the elements put together are not fixed in shape, do not belong to a larger pre-given list but are constructed at least in part as they are entangled together."

8 In Bennett's view, the power of things resides within the process of assemblage (the act of orbiting) as both a property and a prod-

uct—and the continual constituting and re-constituting of assemblage. This gets tricky, however, as typically, for power to be visible, it must materialize through a human being and is thus relational. As in, people still need to recognize a thing's power, regardless if the power originates from the thing itself or the thing's orbit and relation with other things.

9 "Thing-power" essentially is an intervention challenging normative conventions of nature and society as ontologically discrete (Bennett, 2004). Thing-power should not be assumed the only critical theory challenging this divide; it is simply one of many current challenges from a variety of fields and disciplines. Bennett draws upon the works of geochemist Vladimir Ivanovich Vernadsky and philosopher Manuel De Landa, pointing to Vernadsky's seventeenth-century theorization of "living matter" that was incapable of separating life from rocks, mineral from plant; instead, in Verdansky's view, all matter possessed life and power, and was part of ongoing processes (Verdansky did not originate theories of living matter. As Bennet herself writes, others have expanded this idea, including early Greek philosophers and Jains). To ground the power of things and challenge "conventional distinctions between matter and life, inorganic and organic, passive object and active subject," Bennet utilizes De Landa's example of bones: inanimate, more than human, mineral. De Landa observed that bone power (glacier power!) "made new forms of movement control possible among animals, freeing them from many constraints and literally setting them into motion to conquer every available niche in the air, in water, and on land" (Delanda, 1997). Bones enact power that enables people power, animal power, power. Bones are mineralized frames the human body drapes itself upon. Glaciers are shapers of landscapes human society lives within. Can people live without bones (glaciers)? Yes, but what kind of living would that be? What does the distinction amongst 'living' or 'animate' or 'human' matter when thinking about this network of power? To be clear: the power is not in the thing alone—bone power in isolation, in exclusion of cells and soft tissues. Of plants, plastics, and water—all without people. Of just glaciers. The power of these things is in their associations together, in human and more than human communities, not of a thing in a vacuum.

10 People encounter and engage with glacier power in mediated ways, suggesting then that glacier power itself is highly relational. Such relationality may be a critique against the idea of glacier power, but it may also be an expression of its mystery. My intent here is not to fully answer the mediated nature of glacier power. Rather, I make the point that glaciers have and enact power, and that such power must be multidirectional to be realized; there exists no human primacy in glacier power. A mediated experience—e.g., distinguishing glacier power—at some level requires participating with something beyond the human self. Said differently, glacier power indicates a non-human-centric world; that other things (glaciers!) merit consideration. Drawing from Jane Bennet again, participating with other things to some degree affirms that "so-called inanimate things have a life of their own, that deep within them is an inexplicable vitality or energy, a moment of independence from and resistance to us and other things."

11 I lived in Iceland from September 2015 through June 2016. During this time a momentous event occurred in April in Höfn: the release of season six of the HBO show Game of Thrones. While the popular show is filmed throughout the world, many of the scenes are filmed in Iceland, particularly along the southeastern coast with the glaciers. Many Icelanders I met worked in multiple capacities for Game of Thrones during the filming of various seasons of the show. As such, there was high anticipation for the release of the new season—Icelanders wanted to see how their own country was portrayed. I noticed a discernible uptick in Game of Thrones-language in Höfn in reference to the landscape, especially towards glaciers. For example, in Game of Thrones, the fictional land of Westeros and the Seven Kingdoms features a northern border defined by a constructed wall of ice (known simply as the Wall) three hundred miles long and over seven hundred feet tall. As is noted repeatedly in the show, the Wall appears grey or blue depending on the time of day, and may or may not retain some level of magic or sentience as it is thought to have the capacity to defend itself. It is not, by any means, a stretch to relate the Wall to glaciers, and locals regularly referenced the Wall in conversation about any of the region's glaciers.

12 Leifur also told me about once, just before he was going to take the sheep to a higher pasture, a flood came and when he "went to get

the sheep, the water was up to their bellies, and they were keeping the small sheeps on their backs. The sheep were rescuing their own. I had never seen this before."

13 New in the sense that they have not been recently accessed. Many of the pastures were open when the ice was much less in the 1600s.

14 Vatnajökull National Park (VJP) was formerly established as a national park in 2008, but the creation of VJP was a decade in the making primarily based on landownership considerations. In 2002, the Preparatory Committee for the Establishment of Vatnajökull National Park reported that the project (establishment of the park) would most likely go ahead without significant resistance if the park boundaries were delineated along the edge of glaciers. This recommendation was acted upon, and as such, today VJP's boundaries, where possible, follow the 1998 ice margins for the entire ice cap and outlet glaciers. Read here clearly the power of glaciers to settle perceived and existing land ownership conflicts; read also if the park was established on "usable" land/pastureland, conflict might ensue. Instead, the park boundary was placed along the boundary of "non-useable" land.

15 For that matter, the glacier is no longer Kristófer's. Kristófer observed: "When my grandfather bought Holmur in 1952, he bought part of Fláajökull. Well, you know, the plaque about the land, about Hólmur, and then there is a part of Fláajökull included, you know. You own part of the glacier. Part of the glacier was property of Hólmur, until Vatnajökull National Park started a few years ago. So back then, it was my glacier, actually. Not anymore. When the government issued we going to make Vatnajökull glacier National Park, then they basically just took the line where the glacier was under the national park. So now that was government property."

16 While there is substantial evidence that people currently, as in the past, utilize glaciers for mobility, there is also a prevalent narrative that glaciers themselves impact human movement. Glacier change is often framed as a driver for migration, a potential indirect interaction between ice and people. In Nepal, researchers found evidence of temporary migration after glacial lake outburst floods (GLOFS), but all residents returned after a single month. In the glacierized village of Pasu in Pakistan, out-migration was found to be a viable and common practice amongst villagers to diversify household incomes

and not directly linked to glacier change. In Bolivia, glacier recession forcing migration is a powerful and political narrative but without empirical substantiation. Raoul Kaenzig (2015) interviewed 54 people in the Bolivian Andes who migrate regarding their decision-making processes, intentions, and rational. While many migrated due to seasonal water availability concerns (which partially implicates glaciers as direct water resources), migration itself was not triggered per se because of glacier recession.

CHAPTER FIVE

1 Also called dynamic thinning.

2 This phenomenon is described in more detail by local scientist Markús, who said: "What you can actually see here... this is actually what is happening at Heinabergsjökull as well. You know, you know what happened when you have the ice floating? The, it, the behavior of the glacier changes. It starts to, fight back if you can say so. The ice if it doesn't have the friction on the bottom, it is kind of dragged out. ... if you take a coffee or a glass you fill it with water, then you take a cream and if you put it on the rim, it sits on the rim, but the other floats outward. So it is behaving differently than the one still on land. so that is what is actually happening to the glacier... it is losing mass, and it tries to fill up because it is trying to get some sort of balance. ... and it can be explained as the glacier is trying to fill the gap, to reach uh equal, what do you say, a balance. Because when you have this gap, it's not balanced, and it is the same with those glaciers, when it flows up, the ice tongue, it is like it is its pulled towards the other side of the lagoon. Which means, for example, if you look at Heinabergsjökull, he, the uh, the termini has been on the same place more or less for decades. But it is still melting, but you are not aware of it, because it is sort of pulled or dragged forward. It is like it is always trying to fill in the lagoon. But it can't, so it just thinning and thins, and at some point it just starts to break off" (Markús, 2016 #01:05:44-6). Markús's description makes clear that as Heinabergsjökull loosens from the bedrock due to the erosion of the water backfilling underneath it, the edge of the glacier face stretches across the surface of the proglacial lake, causing it to appear year after year as if the edge of the terminus has not changed at all.

3 I visited the glacier three to four times per month between September 2015 and May 2016, and rarely did I encounter anyone else in the area around Heinabergsjökull. Rarely did I meet other visitors. Once, I met a foreign scientist at the glacier lagoon. That scientist was documenting Heinabergsjökull's forelands, which possess unique terracing, kettle holes, push moraines, and gravel outwash fans. As explored in the previous chapter, a local mystery of some debate is whether Heinabergsjökull and eastern neighbor Fláajökull ever joined during or after the Little Ice Age. The scientist I encountered was inspecting Heinabergsjökull's forelands for evidence of the merger, among other tasks, and his previous work correlated with other local work completed—suggesting a positive likelihood for the western merger between Heinabergsjökull and Skálafellsjökull.

4 Helgi Björnsson is a public figure, and as such, I use his real name.

5 Countering a perceived knowledge deficit does not result in people rationally engaging in pro-environmental behaviors. For example, an individual might believe in climate change, and then pursue more knowledge on the topic; conversely, an individual might not believe in climate change, and then not pursue the topic. A focus on just increasing glacier-related knowledge may inevitably increase existing knowledge inequity between people in the community.

6 I use this person's real name as he is a public figure.

CHAPTER SIX

1 While not explicitly focusing on sentience, over the last several decades, researchers have explored and reported on the role and agency of more than human actors (animate and inanimate) (Castree, 2004). Within this work, the human being has progressively been decentralized within a vast entangled network of continually shifting actors (Hinchliffe, 2007, Whatmore, 2002, Latour, 1993), including hybrids (Zimmerer, 2000), cyborgs (Haraway, 2013), things (Bennett, 2004), and many others. Still, though, the idea of sentience within those iterations is largely consigned to Indigenous geographies, or overlooked and unexplored (Brydon, 2012, Lund, 2013.

2 In contrast to dominant articulations of landscape as disconnected from humanity, places described as "dead stage[s] on which humans operate." Landscapes are much more, and to some, quite alive and aware.

3 I do not discuss the glacier Lambatungnajökull much in this book. It is located to the east of Hoffellsjökull, but locals interact less with this glacier in comparison with the others.

4 Philosopher Bruno Latour pointed to Western science's tendency to assume that a single body of knowledge exists, definable and knowable through the Western scientific framework, and it may challenge and defeat (often through processes of "proving") all *other* knowledges (Latour, 2004).

5 Notions of care have been theorized extensively, with scholars examining manifestations of care and emotion in social, culture, and personal practice. Common usage of the term typically revolves around affirmation (e.g. I don't care/I do care), but more broadly use of the term suggests concern for others, human and more than human, tangible and non. Importantly, care also implies simultaneous togetherness and separateness. A caregiver (e.g., a medical professional or social worker) takes care of another (other) person. A separation between the *carer* and the *cared* is discernible. But the act of care bridges separation and produces a sense of cared togetherness.

CHAPTER SEVEN

1 For media reportage of the storm: http://icelandreview.com/news/2016/02/04/storm-wipes-out-travel-plans

2 VJP has monitored visitation to Breiðamerkurjökull via a cable vehicle counter, but the park was not able to monitor the area the entire winter period, nor were park rangers onsite for the entire season. As such, during the winter of 2015-2016, visitation numbers were estimated at 60,000. The winter of 2017-2018, the estimated visitation was around 100,000. The takeaway, however, is that regardless of the exact numer of visitors, the rate of visitation to the area's ice caves in winter has been exceptionally high.

3 Prices vary, but some tour operators in 2016 offered trips to the local ice caves starting at 18,900.00 ISK (~180.00 USD) per person and increasing steeply from there. Some larger companies offered 2-3 departures per day, with 12-15 passengers per vehicle per departure. Companies were regularly fully booked months in advance. As such, ice caves are quite lucrative for companies.

4 Due largely to the Collapse, but also to other factors including the high price of imported building materials, hotels and guesthouses have not been constructed to meet demand. Additionally, new homes for Icelanders have not been constructed. As such, many small communities on the south coast are in dire situations. They need new employees to move to the area, they need families to stay, and they need spaces for tourist accommodation. But the structures are not yet being built. Illustrative of this is the front-page headlines in the Icelandic national newspaper when ground was broken for two new duplexes in Kirkubæjarklaustur, the first homes constructed since 2004 (Hafstad, V., First Homes in Eleven Years. *Reykjavik Grapevine* 2015). Or the January 2016 headlines reporting that most accommodation for tourists across the island was nearly sold out ahead of time for summer 2016 (Hafstad, V., Accommodation Nearly Sold out in Iceland. *Iceland Review* 2016).

5 Building from glaciologist Helgi Björnsson's analogy, if one were to imagine from an aerial perspective the Vatnajökull ice cap as a splayed cat, the head of the beast would be the Öræfajökull volcano in the south, adorned with Iceland's highest peak Hvannadalshnjúkur. The paw directly to the east is Iceland's largest outlet glacier, Skeiðarárjökull, and just to the west of the head is Breiðamerkurjökull. It is the curved Piedmont shape of both glaciers on either side of Hvannadalshnjúkur that evokes the image of a cat resting on two front paws.

6 Breiðamerkurjökull's highest elevation is approximately 6,000 feet, and its lowest point is 130 feet. The equilibrium line, like for all the region's glaciers, has fluctuated extensively, but currently rests just below 4,000 feet.

7 The glacier's lobes flow around the mountains Esjufjöll and Mávabyggðir. As the ice rubs the mountains, this produces extensive debris and wide moraines.

8 As various news outlets reported, Icelandic glaciologist Oddur Sigurðsson told the Iceland Touring Association in March 2016 that the country will likely be denuded of glaciers in less than two hundred years as a direct cause of anthropogenic climate change. Oddur Sigurðsson noted that the glacier melting the most dramatically was Breiðamerkurjökull—and a side effect of the rapid melt has been a relieving of pressure on the earth's mantle, causing Höfn and the surrounding areas to rise by twenty centimeters (7.8 inches) since 1997.

9 Data from 144,000 publicly available images from 2012-2014 tagged with location were mapped by the National Land Survey of Iceland. The map can be found at: http://ferdamalastofa.gistemp.com/vefsjar/vidkomustadir

10 A brief primer on tourism in Iceland: due in part to the isolation associated with Iceland's "Dark Ages" and the positioning of the country on the periphery of Europe, travelers did not seek out the island until late in the eighteenth century. Of those, the primary travelers were European scientists (termed scientist-explorers) like those described in Chapter 2, drawn to the island's glaciers and volcanoes. After WWII, transportation infrastructure established by British and American forces facilitated the development of marginal tourism—people could get around the island using consistent mechanized travel. The tradition of modern day Icelanders utilizing and fetishizing super jeeps stems from post-WWII legacies. As various military forces withdrew or shifted location, they left behind off-road vehicles, jeeps, and other motorized equipment heretofore unseen in Iceland. Icelanders folded this equipment into their lifeways, modifying the vehicles to allow them access to the Icelandic Highlands and various glaciers. Internationally, flights operating from the airfield in Keflavik began carrying passengers in 1944, and in the following decade, Iceland became the main re-fueling stop-over hub for flights between Europe and North America. Intermittent nature-based tourism occurred between 1950-1990, but Iceland was widely considered remote, difficult to access, and unaffordable for average tourists. Tourism did gradually escalate, however, and on the south coast locals developed glacier-based activities as Jökulsárlón rose in popularity.

11 Media worldwide struggled to pronounce the name of the volcano Eyjafjallajökull. Internet memes rapidly ensued. The best compilation of poorly pronounced reportage can be found at www.youtube.com/watch?v=2Q3YVkm8YJM

12 According to Statistics Iceland, 1,767,726 foreign visitors arrived in Iceland in 2016. http://px.hagstofa.is

13 Up from 388,991 people visiting Jökusárlón in 2014.

14 Reports in 2015 state tourism is a 263 billion ISK industry in Iceland.

15 For a 360° panorama of this ice cave, visit: http://www.icecaves.is

16 Ice caves at Kverkjökull are formed primarily through geothermal heat released from volcanic vents in the area; as such, the cave morphology differs significantly from those on the south coast.

17 I would even expand this statement and say the feature is common to all ice, not just glaciers. Arctic sea ice can form into caves under the pressure of tidal shifts. Near Kangiqsujuaq, Greenland, Inuit harvest mussels from ice caves in the Arctic sea ice.

18 Specifically, ice caves form through the process or consequence of moving (dripping, streaming, falling) water, air circulation and convection, or lubrication of deformable wet sediments. Ice caves per se are a generalized label for specific types of glacier features shaped by various processes including moulins, intra-glacial conduits (pipe-like conduits carved into ice), subglacial tunnels (also called subglacial valleys or iceways), sublimated scallops, flutes, and incised meanders, and supraglacial stream channels.

19 Some ice caves flooded over the winter of 2015-2016, and some tourism companies attempted to drain and augment them by using heavy machinery.

20 Some ice caves may be accessed year round, but as a general rule, winter is the safest time to go.

21 Examples of famously unstable ice caves include the Kverkjökull ice caves, which respond to changing geothermal pressures, and also the Paradise Ice Caves on Mt. Rainier in Washington State. These caves were explored in the early 1900s, shifted through the 1940s and became inaccessible, and were located and measured again in the 1970s. The caves vanished entirely in the 1990s.

22 I set up several time-lapse cameras at different points of interaction to understand both who was accessing the ice caves and how the ice caves themselves changed. For example, outside the Waterfall Ice Cave on the eastern-most ice lobe of Breiðamerkurjökull, with Snorri's help we set up a small camera (a GoPro Hero 4) powered by a car battery to take images at one-minute intervals. We set up another camera within the Crystal Ice Cave with the same setup and settings. Surprisingly, the footage revealed not only the clear oscillating rhythms of human traffic, but also the dynamics of how the ice cave flooded during a light rain. Footage can be viewed online at: https://www.youtube.com/watch?v=NVXRtU3Wf3k

23 Only 2% of activities purchased by tourists in 2009 were glacier-based, but by 2012 this was up to 15%. The Icelandic Tourism Board, the national organization overseeing tourism in Iceland, estimates this area of activities might grow by an additional 10% by 2020, suggesting that 25% of tourism activities in Iceland will be associated with glaciers.

24 This person is a public figure and therefore I use his real name.

25 Einar is the great-grandson of Páll Jónsson, the Icelander who in 1891 guided British photographer Frederick W. W. Howell and Þorlákur Þorláksson to the top of the country's tallest peak, Hvannadalshnúkur, achieving what is widely considered to be the first ascent. Einar's father, Sigurður Bjarnason, gave up sheep farming in the early 1990s to repurpose his hay cart to carry travelers and photographers across the tidal sandur out to Ingólfshöfði, a nature reserve for seabirds including puffins, great skuas, and guillemots.

26 The company Local Guide has gone through a re-branding process. It was previously known as From Coast to Mountains, and additionally has an Icelandic branch called Öræfaferðir ehf. Now the companies operate as two distinct entities.

27 In 2016, Einar held the record for the most ascents of Hvannadalshnúkur, though the actual number of ascents varies. I was told numbers between 275 and 305. Whatever the number, the number of ascents is still indicative of Einar's success as a mountain guide.

28 February 27, 2016: 34 people total interviewed (7 women, 27 men). February 27, 2016: 42 people total interviewed (19 women, 23 men). Tourists were primarily from Europe, China, USA, and Iceland.

29 One group of eight tourists had come to do an ice cave tour, but it was canceled.

30 Iceland is not alone. Cultural framings of glaciers regularly position ice as both indicative of climate change and symbolic of climate change. Harkening back to J.B. MacKinnon's (2016) observation in the previous chapter that "glaciers are coming to life, just now, in their twilight hours," it seems that as more ice disappears, the spotlight on the remaining glaciers is growing in global intensity from across social sectors. This in part appears to be related to the overt positioning of glaciers as climate change icons and thus recipients of wide cover-

age in news, films, and other media. The attention has "stimulated increased general interest in them" (Welling, et al., 2015) by tourists worldwide, correlating with a notable increase in visitors to popular glaciated tourist destinations such as Canada's Banff National Park, New Zealand's Westland Tai Poutini National Park, Greenland's Ilulissat Ice Fjord, Argentina's Los Glacier National Park, and, of course, Iceland's south coast. The more glaciers melt, the more people want to see them. While data does not suggest tourists are going to these destinations *solely* due to glaciers, many tourists report that glaciers are in part why they are choosing to visit these destinations.

31 I make this distinction because this situation is likely to change significantly in the coming years.

32 Companies include Glacier Trips, Local Guide, Glacier Adventures, Ice Walk (now Ice Explorers), Ice Guide, Southeast Iceland, and Háfjall.

33 Companies include but are not limited to, Go Ecco, Guide to Iceland, Glacier Guides, Extreme Iceland, and Arctic Adventures. Some companies I interviewed specifically requested they not be named in this research.

34 Adjusted for campgrounds.

35 What Pétur refers to is a bit of poor translation: *jökull* is the Icelandic word for glacier. But English speakers often say: Breiðamerkurjökull Glacier.

36 High clearance vehicles are typically the every-day vehicle for Icelanders, as such many guides do not purchase new vehicles. Only later, when they grow, they might purchase the econ-line vans that transport larger numbers of people.

37 Watching tourists wearing different levels of safety gear in the same space was concerning. Thankfully, often guides from other companies would help "self-police" unsafe practices.

38 For example, if a large foreign school group booked one of the local companies, then many guides would come together to work on this one gig. Full disclosure: I witnessed this exact occurrence numerous times as I helped book groups I was affiliated with.

39 One company, Into the Glacier, constructed a man-made tunnel in

Langjökull, the island's second largest glacier, and offers tours through their structure. Given their close proximity to the country's capital, they attract high numbers of tourists. See: https://intotheglacier.is/tours/

40 During my fieldwork, the Panama Papers were released, and this appeared to rekindle much of the expressed resentment of locals towards those in government and those in the banking industry. For one of the more articulate overviews I have seen on Iceland's Collapse and the subsequent impact of the Panama Papers, see Helgason, H., 600 Silver Lions: How Iceland was Betrayed, Again. *The Reykjavik Grapevine* 2016.

41 In the early years, companies from Reykjavík would book space on local company's trips, often pressuring the operators to add more departures or seats. As Hreiðar, a local man who owns a small guiding company explained, "I was under a lot of pressure from companies in Reykjavík that wanted me to expand, sell more seats on ice cave tours. It started small scale but has really expand. Like, for this winter, I had to open up to selling ice cave tours in May," Hreiðar, 2015.

42 Authorities issued warnings in early April 2016 that the ice caves were unstable and tourists should not purchase tours to them; some companies continued operating and selling tours—local companies in the region stopped.

43 Women historically, in comparison to Europe, have had high rates of participation in the labor market. In 2011, 77% of women engaged in the labor market, of which 64% of women worked full time. In comparison, during the same time, 83.7% of men worked.

44 Many tourists have died at this beach by getting pulled into the ocean. During my fieldwork, two tourists were killed. Signs were posted, but tourists still access the dangerous beach.

CITATIONS

≈ ≈

1. Björnsson, H.; Pálsson, F., Icelandic glaciers. *Jökull* 2008, *58*, 365-386.

2. Sigurðsson, O., Iceland Glaciers. In *Encyclopedia of Snow, Ice and Glaciers*, Springer: 2014; pp 630-636.

3. Björnsson, H., *The Glaciers of Iceland: A Historical, Cultural and Scientific Overview*. Springer: 2016; Vol. 2.

4. Zemp, M.; Roer, I.; Kääb, A.; Hoelzle, M.; Paul, F.; Haeberli, W. *Global glacier changes: facts and figures*; The World Glacier Monitoring Service: The United Nations Environment Programme, 2008.

5. Zemp, M., Gärtner-Roer, I., Nussbaumer, S. U., Hüsler, F., Machguth, H., Mölg, N., Paul, F., and Hoelzle, M., Global Glacier Change Bulletin No. 1 (2012-2013). WGMS, Ed. ICSU(WDS)/ IUGG(IACS)/ UNEP/UNESCO/WMO, World Glacier Monitoring Service: 2015; pp 1-230.

6. Carey, M.; Jackson, M.; Antonello, A.; Rushing, J., Glaciers, gender, and science: A feminist glaciology framework for global environmental change research. *Progress in Human Geography* 2016, 770-793.

7. Orlove, B. S.; Wiegandt, E.; Luckman, B. H., *Darkening peaks: glacier retreat, science, and society*. Univ of California Press: 2008.

8. Jackson, M., Glaciers and climate change: narratives of ruined futures. *Wiley Interdisciplinary Reviews: Climate Change* 2015, *6* (5), 479-492.

9. Pollack, H., *A world without ice*. Penguin: 2010.

10. Adichie, C., The danger of a single story. *TED Ideas worth spreading* 2009.

11. Greschke, H., The Social Facts of Climate Change: An Ethnographic Approach. In *Grounding Global Climate Change*, Springer: 2015; pp 121-138.

12. Smith, P.; Howe, N., *Climate Change as Social Drama: Global Warming in the Public Sphere*. Cambridge University Press: 2015.

13. Marshall, G., *Don't even think about it: Why our brains are wired to ignore climate change*. Bloomsbury Publishing USA: 2015.

14. Hulme, M., *Why we disagree about climate change*. Cambridge University Press: 2009.

15. Sörlin, S., Ice diplomacy and climate change: Hans Ahlmann and the quest for a nordic region beyond borders. *Science, geopolitics and culture in the polar region–Norden beyond borders* 2013, 23-54.

16. Adger, W. N.; Barnett, J.; Brown, K.; Marshall, N.; O'Brien, K., Cultural dimensions of climate change impacts and adaptation. *Nature Climate Change* 2013, *3* (2), 112-117.

17. Carey, M., *In the Shadow of Melting Glaciers: Climate Change and Andean Society*. Oxford University Press: New York, 2010.

18. Cruikshank, J., *Do Glaciers Listen? Local Knowledge, Colonial Encounters, and Social Imagination*. UBC Press: Vancouver, Canada, 2005.

19. Cruikshank, J., Are Glaciers 'Good to Think With'? Recognising Indigenous Environmental Knowledge. *Anthropological Forum* 2012, *22* (3), 239-250.

20. Cruikshank, J., Encountering Glaciers: Two Centuries of Stories from the Saint Elias Mountains, Northwestern North America. In *Landscapes Beyond Land: Routes, Aesthetics, Narratives*, Árnason, A.; Ellison, N.; Vergunst, J.; Whitehouse, A., Eds. Berghahn Books: New York, 2012; pp 49-66.

21. Allison, E. A., The spiritual significance of glaciers in an age of climate change. *Wiley Interdisciplinary Reviews: Climate Change* 2015, *6* (5), 493-508.

22. Drew, G., A retreating goddess? Conflicting perceptions of ecological change near the Gangotri-Gaumukh Glacier. *Journal for the Study of Religion, Nature and Culture* 2012, *6* (3), 344-362.

23. Drew, G., Why wouldn't we cry? Love and loss along a river in decline. *Emotion, Space and Society* 2013, *6*, 25-32.

24. Welling, J. T.; Árnason, Þ.; Ólafsdottír, R., Glacier tourism: a scoping review. *Tourism Geographies* 2015, *17* (5), 635-662.

25. Sörlin, S., Cryo-History: Narratives of Ice and the Emerging Arctic Humanities. In *The New Arctic*, Springer: 2015; pp 327-339.

26. Kaijser, A., White ponchos dripping away? *Interpretive Approaches to Global Climate Governance:(De) constructing the Greenhouse* 2013, 183.

27. Gore, A., Foreword. In *A World Without Ice*, Pollack, H., Ed. Avery: New York, 2010.

28. Wong, E., Chinese Glacier's Retreat Signals Trouble for Asian Water Supply. *The New York Times* 2015.

29. Schama, S., *Landscape and Memory*. Random House: Toronto, 1995.

30. Guðmundsson, A.; Kjartansson, H.; Douglas, G., *Living earth: outline of the geology of Iceland*. Mál og menning: 2007.

31. Thordarson, T.; Hoskuldsson, A., *Iceland: Classic Geology in Europe*. Second ed.; Dunedin Academic PressLtd: 2002.

32. Larsen, G.; Gudmundsson, M. T.; Björnsson, H., Eight centuries of periodic volcanism at the center of the Iceland hotspot revealed by glacier tephrostratigraphy. *Geology* 1998, *26* (10), 943-946.

33. Sigurðsson, O.; Williams, R. S. *Geographic names of Iceland's glaciers: historic and modern*; 0607978155; US Department of the Interior, US Geological Survey: 2008.

34. Hambrey, M. J.; Alean, J. C., Colour Atlas of Glacial Phenomena. CRC Press: 2016.

35. Pfeffer, W. T.; Arendt, A. A.; Bliss, A.; Bolch, T.; Cogley, J. G.; Gardner, A. S.; Hagen, J.-O.; Hock, R.; Kaser, G.; Kienholz, C., The Randolph Glacier Inventory: a globally complete inventory of glaciers. *Journal of Glaciology* 2014, *60* (221), 537-552.

36. Cogley, J.; Hock, R.; Rasmussen, L.; Arendt, A.; Bauder, A.; Braithwaite, R.; Jansson, P.; Kaser, G.; Möller, M.; Nicholson, L., Glossary of glacier mass balance and related terms, IHP-VII technical documents in hydrology No. 86, IACS Contribution No. 2. *International Hydrological Program, UNESCO, Paris* 2011.

37. Crochet, P.; Jóhannesson, T.; Jónsson, T.; Sigurðsson, O.; Björnsson, H.; Pálsson, F.; Barstad, I., Estimating the spatial distribution of precipitation in Iceland using a linear model of orographic precipitation. *Journal of Hydrometeorology* 2007, *8* (6), 1285-1306.

38. Rögnvaldsson, Ó.; Jónsdóttir, J. F.; Ólafsson, H., Numerical simulations of precipitation in the complex terrain of Iceland–Comparison with glaciological and hydrological data. *Meteorologische Zeitschrift* 2007, *16* (1), 71-85.

39. Björnsson, H., *Jöklar á íslandi*. Bókaútgáfan Opna: 2009.

40. Hannesdóttir, H.; Björnsson, H.; Pálsson, F.; Aðalgeirsdóttir, G.; Guðmundsson, S., Changes in the southeast Vatnajökull ice cap, Iceland, between~ 1890 and 2010. *The Cryosphere* 2015, *9* (2), 565-585.

41. Hannesdóttir, H.; Björnsson, H.; Pálsson, F.; Aðalgeirsdóttir, G.; Guðmundsson, S., Variations of southeast Vatnajökull ice cap (Iceland) 1650–1900 and reconstruction of the glacier surface geometry at the Little Ice Age maximum. *Geografiska Annaler: Series A, Physical Geography* 2015, *97* (2), 237-264.

42. Einarsson, T.; Albertsson, K. J., The glacial history of Iceland during the past three million years. *Philosophical Transactions of the Royal Society of London B: Biological Sciences* 1988, *318* (1191), 637-644.

43. Imbrie, J.; Boyle, E.; Clemens, S.; Duffy, A.; Howard, W.; Kukla, G.; Kutzbach, J.; Martinson, D.; McIntyre, A.; Mix, A., On the structure and origin of major glaciation cycles 1. Linear responses to Milankovitch forcing. *Paleoceanography* 1992, *7* (6), 701-738.

44. Ingólfsson, Ó.; Norðdahl, H.; Schomacker, A., 4 Deglaciation and Holocene Glacial History of Iceland. *Developments in Quaternary Sciences* 2010, *13*, 51-68.

45. Allen, J.; Cochrane, A., Beyond the territorial fix: regional assemblages, politics and power. *Regional studies* 2007, *41* (9), 1161-1175.

46. Hannesdóttir, H. Variations of southeast Vatnajökull, past, present and future. University of Iceland, Reykjavik, 2014.

47. Kellogg, R.; Smiley, J., *The Sagas of Icelanders (Penguin Classics Deluxe Edition).* Penguin: New York, 2001.

48. Karlsson, G.; Pálsson, G., From Sagas to Society: Comparative Approaches to Early Iceland. JSTOR: 1996.

49. Karlsson, G., *The history of Iceland.* U of Minnesota Press: 2000.

50. Price, T. D.; Gestsdóttir, H., The first settlers of Iceland: an isotopic approach to colonisation. *Antiquity* 2006, *80* (307), 130-144.

51. Vésteinsson, O., Patterns of settlement in Iceland: a study in prehistory. *Saga book-Viking Society for Northern Research* 1998, *25*, 1-29.

52. Hastrup, K., Icelandic topography and the sense of identity. *Nordic landscapes* 2008, 53-76.

53. Oslund, K., *Iceland imagined: Nature, culture, and storytelling in the North Atlantic.* University of Washington Press: 2011.

54. Quinn, J., From orality to literacy in medieval Iceland. *CAMBRIDGE STUDIES IN MEDIEVAL LITERATURE* 2000, *42*, 30-60.

55. Pálsdóttir, I. H., Promoting Iceland: The shift from nature to people's power. *Place Branding and Public Diplomacy* 2016, *12* (2-3), 210-217.

56. Ives, J. D., *Skaftafell in Iceland: a thousand years of change.* Ormstunga: 2007.

57. Hastrup, K., *Nature and policy in Iceland 1400-1800: an anthropological analysis of history and mentality.* Clarendon Press: 1990.

58. Pálsson, S., *Draft of a physical, geographical, and historical description of Icelandic ice mountains on the basis of a journey to the most*

prominent of them in 1792-1794 with four maps and eight perspective drawings. Icelandic Literary Society: Reykjavik, 2004.

59. Byock, J. L., History and the sagas: the effect of nationalism. *From Sagas to society: Comparative approaches to early Iceland* 1992, 43-59.

60. Huijbens, E. H.; Benediktsson, K., Practising highland heterotopias: Automobility in the interior of Iceland. *Mobilities* 2007, *2* (1), 143-165.

61. Júlíusdóttir, M.; Skaptadóttir, U. D.; Karlsdóttir, A., Gendered migration in turbulent times in Iceland. *Norsk Geografisk Tidsskrift-Norwegian Journal of Geography* 2013, *67* (5), 266-275.

62. Bjarnason, T.; Thorlindsson, T., Should I stay or should I go? Migration expectations among youth in Icelandic fishing and farming communities. *Journal of Rural Studies* 2006, *22* (3), 290-300.

63. Karlsdóttir, A.; Ingólfsdóttir, A. H., Gendered Outcomes of Socio-economic Restructuring: A Tale from a Rural Village in Iceland. *NORA-Nordic Journal of Feminist and Gender Research* 2011, *19* (3), 163-180.

64. Eðvarðsson, I. R.; Heikkilä, E.; Johansson, M.; Johannesson, H.; Rauhut, D.; Schmidt, T. D.; Stamböl, L. S.; Wilkman, S., *Demographic Changes, Labour Migration and EU-enlargement*. Nordregio: 2007.

65. Skaptadottir, U. D., Responses to global transformations: Gender and ethnicity in resource-based localities in Iceland. *Polar Record* 2004, *40* (03), 261-267.

66. Sigurðardóttir, H. R. Svavar Guðnason listmálari og menningarverðmætin: Sýning um Svavar Guðnason á Höfn í Hornafirði. University of Iceland, 2010.

67. Aðalgeirsdóttir, G.; Guðmundsson, S.; Björnsson, H.; Pálsson, F.; Jóhannesson, T.; Hannesdóttir, H.; Sigurdsson, S.; Berthier, E., Modelling the 20th and 21st century evolution of Hoffellsjökull glacier, SE-Vatnajökull, Iceland. *The Cryosphere* 2011, *5* (4), 961-975.

68. Finnbogadóttir, V., Foreword. In *The Glaciers of Iceland: A Historical, Cultural and Scientific Overview*, Björnsson, H., Ed. Springer: 2016; Vol. 2.

69. Grove, J. M., *The Little Ice Age*. Routledge: New York, 1988.

70. Þórarinsson, S., *Glaciological knowledge in Iceland before 1800: A historical outline*. 1960.

71. Sigurðsson, O.; Williams, R. S., Geographic Names of Iceland's Glaciers: Historic and Modern. Island, N. E. A., Ed. US Department of the Interior, US Geological Survey: US Department of the Interior, US Geological Survey, 2008.

72. Ingolfsson, O., 'On glaciers in general and particular...': The life and works of an Icelandic pioneer in glacial research. *Boreas* 1991, *20* (1), 79-84.

73. Richard S. Williams, J.; Sigurðsson, O., Introduction. In Sveinn Pálsson's Draft of a physical, geographical, and historical description of Icelandic ice mountains on the basis of a journey to the most prominent of them in 1792-1794 with four maps and eight perspective drawings. Icelandic Literary Society: Reykjavik, 2004.

74. Sörlin, S., Hans W: son Ahlmann, Arctic research and polar warming. *Reprint/Northern Studies Programme, Umeå Univ., 14* 1997, 383-398.

75. Sörlin, S., Narratives and counter-narratives of climate change: North Atlantic glaciology and meteorology, c. 1930–1955. *Journal of Historical Geography* 2009, *35* (2), 237-255.

76. Sörlin, S., The Anxieties of a Science Diplomat: Field Coproduction of Climate Knowledge and the Rise and Fall of Hans Ahlmann's "Polar Warming". *Osiris* 2011, *26* (1), 66-88.

77. McKinzey, K. M.; Olafsdóttir, R.; Dugmore, A. J., Perception, history, and science: coherence or disparity in the timing of the Little Ice Age maximum in southeast Iceland? *Polar Record* 2005, *41* (04), 319-334.

78. Deutscher, G., *Through the language glass: Why the world looks different in other languages*. Metropolitan Books: 2010.

79. Solnit, R., *A field guide to getting lost*. Canongate Books: 2006.

80. Grímsson, Ó. R., An Address by the President of Iceland Ólafur Ragnar Grímsson at the opening of the Kjarval Exhibition The State Russian Museum, St. Petersburg. Iceland, O. o. t. P. o., Ed. 2013.

81. Bolin, I., The glaciers of the Andes are melting: indigenous and anthropological knowledge merge in restoring water resources. *Anthropology & Climate Change. From Encounters to Actions* 2009, 228-39.

82. Carreño, G. S., The Glacier, the Rock, the Image: Emotional Experience and Semiotic Diversity at the Quyllurit'i Pilgrimage (Cuzco, Peru). *Signs and Society* 2014, *2* (S1), S188-S214.

83. Olsen, M.; Callaghan, T.; Reist, J.; Reiersen, L.; Dahl-Jensen, D.; Granskog, M.; Goodison, B.; Hovelsrud, G.; Johansson, M.; Kallenborn, R., The changing Arctic cryosphere and likely consequences: An overview. *Ambio* 2011, *40* (1), 111-118.

84. Cruikshank, J., Glaciers and climate change: Perspectives from oral tradition. *Arctic* 2001, 377-393.

85. Brugger, J.; Dunbar, K.; Jurt, C.; Orlove, B., Climates of anxiety: Comparing experience of glacier retreat across three mountain regions. *Emotion, Space and Society* 2013, *6*, 4-13.

86. Orlove, B. S.; Strauss, S., *Weather, climate, culture*. Berg: 2003.

87. Mahamane, M.; Hochschild, V.; Schultz, A.; Kuma, J., Monitoring Desertification in the Tillabéry Landscape (Sahel Region) using Change Detection Methods and Landscape Metrics. *IJAR* 2015, *1* (10), 315-321.

88. Uddin, M. N.; Bokelmann, W.; Entsminger, J. S., Factors affecting farmers' adaptation strategies to environmental degradation and climate change effects: a farm level study in Bangladesh. *Climate* 2014, *2* (4), 223-241.

89. Clarke, D.; Williams, S.; Jahiruddin, M.; Parks, K.; Salehin, M., Projections of on-farm salinity in coastal Bangladesh. *Environmental Science: Processes & Impacts* 2015, *17* (6), 1127-1136.

90. Björnsson, H., Subglacial lakes and jökulhlaups in Iceland. *Global and Planetary Change* 2003, *35* (3), 255-271.

91. Gudmundsson, M. T.; Sigmundsson, F.; Björnsson, H., Ice–volcano interaction of the 1996 Gjálp subglacial eruption, Vatnajökull, Iceland. *Nature* 1997, *389* (6654), 954-957.

92. Miège, C.; Forster, R. R.; Brucker, L.; Koenig, L. S.; Solomon, D. K.; Paden, J. D.; Box, J. E.; Burgess, E. W.; Miller, J. Z.; McNerney, L., Spatial extent and temporal variability of Greenland firn aquifers detected by ground and airborne radars. *Journal of Geophysical Research: Earth Surface* 2016, *121* (12), 2381-2398.

93. Nye, J., Water flow in glaciers: jökulhlaups, tunnels and veins. *Journal of Glaciology* 1976, *17* (76), 181-207.

94. Jónsson, S. A.; Benediktsson, Í. Ö.; Ingólfsson, Ó.; Schomacker, A.; Bergsdóttir, H. L.; Jacobson, W. R.; Linderson, H., Submarginal drumlin formation and late Holocene history of Fláajökull, southeast Iceland. *Annals of Glaciology* 2016, 1-14.

95. Tweed, F. S.; Carrivick, J. L., Deglaciation and proglacial lakes. *Geology Today* 2015, *31* (3), 96-102.

96. Dabski, M., Dating of the Flàajökull moraine ridges SE Iceland; comparison of the glaciological, cartographic and lichenometrical data. *Jökull* 2002, *51*, 17-24.

97. Evans, D. J.; Ewertowski, M.; Orton, C., Fláajökull (north lobe), Iceland: active temperate piedmont lobe glacial landsystem. *Journal of Maps* 2015, 1-13.

98. Eiríksson, H. H., *Observations and measurements of some Glaciers in Austur-Skaftafellssýsla*. Gutenberg: 1932.

99. Corbridge, S., *Seeing the state: Governance and governmentality in India*. Cambridge University Press: 2005; Vol. 10.

100. Hall, S., Foucault: Power, knowledge and discourse. *Discourse theory and practice: A reader* 2001, *72*, 81.

101. Steven, L., Power: A radical view. *London and New York: Macmillan* 1974.

102. Law, J., *After method: Mess in social science research*. Routledge: 2004.

103. Deleuze, G.; Guattari, F., *A thousand plateaus: Capitalism and schizophrenia*. Bloomsbury Publishing: 1988.

104. Mitchell, T., *Rule of experts: Egypt, techno-politics, modernity*. Univ of California Press: 2002.

105. Nash, L., The agency of nature or the nature of agency? *Environmental History* 2005, *10* (1), 67-69.

106. Stewart, M. A., *What nature suffers to groe: life, labor, and landscape on the Georgia coast, 1680-1920*. University of Georgia Press: 2002; Vol. 19.

107. Tsing, A., *The Mushroom at the End of the World: On the Possibility of Life in Capitalist Ruins*. Princeton University Press: 2015.

108. Bennett, J., The force of things steps toward an ecology of matter. *Political theory* 2004, *32* (3), 347-372.

109. Bennett, J., The Force of Things: Steps Toward an Ecology of Matter. *Political theory* 2004, *32* (3), 347-372.

110. Bradatan, C., Scaling the 'Wall in the Head'. *New York Times Blogs, Opinionator* 2011.

111. Macfarlane, R., *Mountains of the Mind*. Granta: 2009.

112. Marshall, G., *Don't even think about it: Why our brains are wired to ignore climate change*. Bloomsbury Publishing USA: 2015.

113. Evans, D. J.; Orton, C., Heinabergsjökull and Skalafellsjökull, Iceland: active temperate piedmont lobe and outwash head glacial landsystem. *Journal of Maps* 2015, *11* (3), 415-431.

114. Corry, O.; Jørgensen, D., Beyond 'deniers' and 'believers': Towards a map of the politics of climate change. *Global Environmental Change* 2015, *32*, 165-174.

115. Reser, J. P.; Bradley, G. L.; Ellul, M. C., Encountering climate change: 'seeing' is more than 'believing'. *Wiley Interdisciplinary Reviews: Climate Change* 2014, *5* (4), 521-537.

116. Stocking, S. H., On drawing attention to ignorance. *Science Communication* 1998, *20* (1), 165-178.

117. Gross, M., The unknown in process dynamic connections of ignorance, non-knowledge and related concepts. *Current Sociology* 2007, *55* (5), 742-759.

118. Kempner, J.; Merz, J. F.; Bosk, C. L. In *Forbidden Knowledge: Public Controversy and the Production of Nonknowledge1*, Sociological forum, Wiley Online Library: 2011; pp 475-500.

119. Kempner, J.; Perlis, C. S.; Merz, J. F., Forbidden knowledge. *Science* 2005, *307* (5711), 854-854.

120. Stern, P. C., New environmental theories: toward a coherent theory of environmentally significant behavior. *Journal of social issues* 2000, *56* (3), 407-424.

121. Gardner, G. T.; Stern, P. C., *Environmental problems and human behavior*. Allyn & Bacon: 1996.

122. Kollmuss, A.; Agyeman, J., Mind the gap: why do people act environmentally and what are the barriers to pro-environmental behavior? *Environmental education research* 2002, *8* (3), 239-260.

123. Dunlap, R. E.; McCright, A. M., A widening gap: Republican and Democratic views on climate change. *Environment: Science and Policy for Sustainable Development* 2008, *50* (5), 26-35.

124. Kaijser, A., 12 White ponchos dripping away? *Interpretive Approaches to Global Climate Governance:(De) constructing the Greenhouse* 2013, 183.

125. Dunbar, K. W.; Brugger, J.; Jurt, C.; Orlove, B. S., Comparing Knowledge of and Experience with Climate Change Across Three Glaciated Mountain Regions. *Climate Change and Threatened Communities ; Vulnerability, Capacity, and Action* 2012, 93-106.

126. Rayner, S., Uncomfortable knowledge: the social construction of ignorance in science and environmental policy discourses. *Economy and Society* 2012, *41* (1), 107-125.

127. O'Brien, K., Global environmental change III Closing the gap between knowledge and action. *Progress in Human Geography* 2013, *37* (4), 587-596.

128. Norgaard, K. M., *Living in denial: Climate change, emotions, and everyday life*. MIT Press: 2011.

129. Cohen, S., *States of denial: Knowing about atrocities and suffering*. John Wiley & Sons: 2013.

130. Stanley, C., *States of Denial: Knowing about Atrocities and Suffering.* John Wiley & Sons: 2001.

131. Editorial, Nýtt kuldaskeið gæti tekið við. *Morgunblaðið* 2015.

132. Drew, G., Ecological Change and the Sociocultural Consequences of the Ganges River's Decline. 2011, 203-218.

133. Drew, G. 'Ganga is Disappearing': Women, Development, and Contentious Practice on the Ganges River. The University Of North Carolina at Chapel Hill, 2011.

134. Pálsson, G., These are not old ruins: A heritage of the Hrun. *International Journal of Historical Archaeology* 2012, *16* (3), 559-576.

135. Aldred, O.; Lucas, G., Events, Temporalities, and Landscapes in Iceland. *Bolender* 2010, *2010*, 189-198.

136. Kuran, T., The unthinkable and the unthought. *Rationality and Society* 1993, *5* (4), 473-505.

137. Mitzen, J., Security Communities and the Unthinkabilities of War. *Annual Review of Political Science* 2016, *19*, 229-248.

138. Ossorio, P. G., *The Behavior of Persons.* Descriptive Psychology Press: 2006.

139. Edkins, J., *Trauma and the Memory of Politics.* Cambridge University Press: 2003.

140. MacKinnon, J. B., The Whale Dying on the Mountain. *Hakai Magazine* 2/16/2016, 2016.

141. Milman, O., Greenland's ice melt accelerating as surface darkens, raising sea levels. *The Guardian* 2016.

142. USGS Repeat Photography of Alaskan Glaciers. https://www2.usgs.gov/climate_landuse/glaciers/repeat_photography.asp (accessed January 21, 2017).

143. Cullen, N.; Sirguey, P.; Mölg, T.; Kaser, G.; Winkler, M.; Fitzsimons, S., A century of ice retreat on Kilimanjaro: the mapping reloaded. *The Cryosphere* 2013, *7* (2), 419-431.

144. Pepin, N.; Duane, W.; Schaefer, M.; Pike, G.; Hardy, D., Measuring and modeling the retreat of the summit ice fields on Kilimanjaro, East Africa. *Arctic, Antarctic, and Alpine Research* 2014, *46* (4), 905-917.

145. Kunzig, R., Meltdown. *National Geographic* 2013.

146. Jurt, C.; Burga, M. D.; Vicuña, L.; Huggel, C.; Orlove, B. S., Local perceptions in climate change debates: insights from case studies in the Alps and the Andes. *Climatic Change* 2015, *133* (3), 511-523.

147. Hewitt, K., Glaciers in Human Life. In *Glaciers of the Karakoram Himalaya*, Springer: 2014; pp 327-351.

148. Hay, J.; Elliott, T., New Zealand's glaciers. In *Darkening peaks: Glacier retreat, science, and society*, Orlove, B.; Wiegandt, E.; Luckman, B., Eds. University of California Press: Berkeley, 2008; pp 185-195.

149. Grønnow, B., Blessings and horrors of the interior: Ethno-historical studies of Inuit perceptions concerning the inland region of west Greenland. *Arctic Anthropology* 2009, *46* (1-2), 191-201.

150. Ingold, T., Rethinking the animate, re-animating thought. *Ethnos* 2006, *71* (1), 9-20.

151. Wells, P., *Understanding Animation*. Psychology Press: 1998.

152. Harvey, G., *Animism: Respecting the living world*. Wakefield Press: 2005.

153. Gergan, M. D., Animating the Sacred, Sentient and Spiritual in Post-Humanist and Material Geographies. *Geography Compass* 2015, *9* (5), 262-275.

154. Hsu, M.; Howitt, R.; Chi, C. C., The idea of 'Country': Reframing post-disaster recovery in Indigenous Taiwan settings. *Asia Pacific Viewpoint* 2014, *55* (3), 370-380.

155. Peterson, N., Is the Aboriginal landscape sentient? Animism, the new animism and the Warlpiri. *Oceania* 2011, *81* (2), 167.

156. Garuba, H., On animism, modernity/colonialism and the African order of knowledge: Provisional reflections. *Contested Ecologies* 2013, 42.

157. Costa, L.; Fausto, C., The return of the animists: Recent studies of Amazonian ontologies. *Religion and Society: Advances in Research* 2010, *1* (1), 89-109.

158. Brightman, M.; Grotti, V. E.; Ulturgasheva, O., *Animism in rainforest and tundra: personhood, animals, plants and things in contemporary Amazonia and Siberia*. Berghahn Books: 2014.

159. Pedersen, M. A., Totemism, animism and North Asian indigenous ontologies. *Journal of the Royal Anthropological institute* 2001, *7* (3), 411-427.

160. Roncoli, C.; Crane, T.; Orlove, B., Fielding climate change in cultural anthropology. Left Coast Press, San Francisco, CA: 2009; pp 87-115.

161. Brydon, A., Sentience. *Conversations with Landscape* 2012, 193-209.

162. Nadasdy, P., *Hunters and bureaucrats: power, knowledge, and aboriginal-state relations in the southwest Yukon*. UBC Press: 2004.

163. Krupnik, I.; Jolly, D., *The Earth Is Faster Now: Indigenous Observations of Arctic Environmental Change. Frontiers in Polar Social Science*. ERIC: 2002.

164. Laidler, G. J., Inuit and scientific perspectives on the relationship between sea ice and climate change: the ideal complement? *Climatic Change* 2006, *78* (2-4), 407-444.

165. Oreskes, N., Science and public policy: what's proof got to do with it? *Environmental Science & Policy* 2004, *7* (5), 369-383.

166. Latour, B., Whose cosmos, which cosmopolitics? Comments on the peace terms of Ulrich Beck. *Common Knowledge* 2004, *10* (3), 450-462.

167. Tsing, A.; Satsuka, S., Diverging understandings of forest management in matsutake science. *Economic Botany* 2008, *62* (3), 244-253.

168. Tsing, A., Beyond economic and ecological standardisation. *The Australian Journal of Anthropology* 2009, *20* (3), 347-368.

169. Tsing, A., Unruly edges: mushrooms as companion species. *Environmental Humanities* 2012, *1*, 141-154.

170. Tsing, A., Sorting out commodities: How capitalist value is made through gifts. *HAU: Journal of Ethnographic Theory* 2013, *3* (1), 21-43.

171. Grímsson, Ó. R., The New Importance of the Arctic and the Himalayas. In *Speech*, Chicago Council on Global Affairs: Ice and Water in the 21st Century, 2014.

172. Head, L.; Gibson, C., Becoming differently modern Geographic contributions to a generative climate politics. *Progress in Human Geography* 2012, *36* (6), 699-714.

173. Ólafsson, S., Iceland's financial crisis and level of living consequences. 2011.

174. Hall, C. M.; Saarinen, J.; Hall, C.; Saarinen, J., Last chance to see? Future issues for polar tourism and change. *Tourism and Change in Polar Regions: Climate, Environments and Experiences. London: Routledge* 2010.

175. Lemelin, H.; Dawson, J.; Stewart, E. J.; Maher, P.; Lueck, M., Last-chance tourism: The boom, doom, and gloom of visiting vanishing destinations. *Current Issues in Tourism* 2010, *13* (5), 477-493.

176. Lemelin, H.; Dawson, J.; Stewart, E. J., *Last chance tourism: Adapting tourism opportunities in a changing world.* Routledge: 2013.

177. Müller, D. K.; Lundmark, L.; Lemelin, R. H., Introduction: New issues in polar tourism. In *New issues in polar tourism*, Springer: 2013; pp 1-17.

178. Hulme, M., Reducing the future to climate: A story of climate determinism and reductionism. *Osiris* 2011, *26* (1), 245-266.

179. Swyngedouw, E., Apocalypse forever? Post-political populism and the spectre of climate change. *Theory, Culture & Society* 2010, *27* (2-3), 213-232.

180. Demeritt, D., Geography and the promise of integrative environmental research. *Geoforum* 2009, *40* (2), 127-129.

181. Liverman, D., Conventions of climate change: constructions of danger and the dispossession of the atmosphere. *Journal of Historical Geography* 2009, *35* (2), 279-296.

182. Storrar, R. D.; Evans, D. J.; Stokes, C. R.; Ewertowski, M., Controls on the location, morphology and evolution of complex esker systems at decadal timescales, Breiðamerkurjökull, southeast Iceland. *Earth Surface Processes and Landforms* 2015, *40* (11), 1421-1438.

183. Canas, D.; Chan, W. M.; Chiu, A.; Jung-Ritchie, L.; Leung, M.; Pillay, L.; Waltham, B., Potential Environmental Effects of Expanding Lake Jökulsárlón in Response to Melting of Breiðamerkurjökull, Iceland. *Cartographica: The International Journal for Geographic Information and Geovisualization* 2015, *50* (3), 204-213.

184. Ahmed, N., White House warned on imminent Arctic ice death spiral. *The Guardian* 2013.

185. Mooney, C., "The melting of Antarctica was already really bad. It just got worse". *The Washington Post* 2015.

186. McKie, R., "Polar melt down see us on an icy road to disaster". *The Guardian* 2015.

187. Mooney, C., "As Greenland melts, this iconic glacier is creating terrifying tsunamis". *The Washington Post* 2016.

188. Gagné, K.; Rasmussen, M. B.; Orlove, B. S., Glaciers and society: attributions, perceptions, and valuations. *Wiley Interdisciplinary Reviews: Climate Change* 2014, *5* (6), 793-808.

189. Orlove, B. S., Glacier retreat: reviewing the limits of human adaptation to climate change. *Environment: Science and Policy for Sustainable Development* 2009, *51* (3), 22-34.

190. Rhoades, R. E.; Zapata Ríos, X.; Aragundy Ochoa, J., Mama Cotacachi: history, local perceptions, and social impacts of climate change and glacier retreat in the Ecuadorian Andes. *In: Orlove, B., E. Wiegandt and BH Luckman (eds.). Darkening Peaks: Glacier Retreat, Science, and Society, 216-225* 2016.

191. Hastrup, K., Comparing Climate Worlds: Theorising across Ethnographic Fields. In *Grounding Global Climate Change*, Springer: 2015; pp 139-154.

192. O'Brien, K. L.; Leichenko, R. M., Winners and losers in the context of global change. *Annals of the association of American geographers* 2003, *93* (1), 89-103.

193. Compton, K.; Bennett, R. A.; Hreinsdóttir, S., Climate-driven vertical acceleration of Icelandic crust measured by continuous GPS geodesy. *Geophysical Research Letters* 2015, *42* (3), 743-750.

194. Xavier, J. C.; Peck, L. S.; Fretwell, P.; Turner, J., Climate change and polar range expansions: Could cuttlefish cross the Arctic? *Marine Biology* 2016, *163* (4), 1-5.

195. Björnsson, H.; Jóhannesson, T.; Snorrason, Á., Recent climate change, projected impacts, and adaptation capacity in Iceland. In *Climate*, Springer: 2011; pp 465-475.

196. Saba, V. S.; Griffies, S. M.; Anderson, W. G.; Winton, M.; Alexander, M. A.; Delworth, T. L.; Hare, J. A.; Harrison, M. J.; Rosati, A.; Vecchi, G. A., Enhanced warming of the Northwest Atlantic Ocean under climate change. *Journal of Geophysical Research: Oceans* 2016, *121* (1), 118-132.

197. Wells, N. C., The North Atlantic Ocean and climate change in the UK and northern Europe. *Weather* 2016, *71* (1), 3-6.

198. Benediktsson, K.; Lund, K. A.; Huijbens, E., Inspired by eruptions? Eyjafjallajökull and Icelandic tourism. *Mobilities* 2011, *6* (1), 77-84.

199. Hafstad, V., Iceland Expects 1.73 Million Tourists. *Iceland Review* 2016.

200. Óladóttir, O. Þ. *Tourism in Iceland in Figures*; Icelandic Tourist Board: Akureyri, 2016.

201. Þórhallsdóttir, G.; Ólafsson, R. *Fjöldi gesta í Vatnajökulsþjóðgarði [Number of visitors in Vatnajokull National Park]*; University of Iceland: Vatnajökulsþjóðgarður, Klapparstíg 25-27, IS-101 Reykjavík, 2015.

202. Thorsteinsson, P. *Inbound tourism expenditure was 263 billion ISK in 2015*; Statistics Iceland Reykjavik 2017.

203. Welling, J.; Árnason, T., External and Internal Challenges of Glacier Tourism Development in Iceland. *Mountain Tourism: Experiences, Communities, Environments and Sustainable Futures* 2016, 174.

204. Wilson, J.; Espiner, S.; Stewart, E.; Purdie, H., *'Last chance tourism'at the Franz Josef and Fox Glaciers, Westland Tai Poutini National Park: A survey of visitor experience.* LEaP: 2014.

205. Gulley, J.; Benn, D.; Müller, D.; Luckman, A., A cut-and-closure origin for englacial conduits in uncrevassed regions of polythermal glaciers. *Journal of Glaciology* 2009, *55* (189), 66-80.

206. Gulley, J.; Benn, D.; Screaton, E.; Martin, J., Mechanisms of englacial conduit formation and their implications for subglacial recharge. *Quaternary Science Reviews* 2009, *28* (19), 1984-1999.

207. Strauss, S., Are cultures endangered by climate change? Yes, but.... *Wiley Interdisciplinary Reviews: Climate Change* 2012, *3* (4), 371-377.

208. Devlin, H., Receding glacier causes immense Canadian river to vanish in four days. *The Guardian* 2017.

209. King, T., *The truth about stories: A native narrative*. House of Anansi: 2003.

210. Piersall, A.; Halvorson, S. J., Local perceptions of glacial retreat and livelihood impacts in the At-Bashy Range of Kyrgyzstan. *GeoJournal* 2014, *79* (6), 693-703.

211. Mangerud, J.; Gosse, J.; Matiouchkov, A.; Dolvik, T., Glaciers in the Polar Urals, Russia, were not much larger during the Last Global Glacial Maximum than today. *Quaternary Science Reviews* 2008, *27* (9), 1047-1057.

212. Nuttall, M., Living in a world of movement: human resilience to environmental instability in Greenland. *Anthropology and climate change: from encounters to actions* 2009, 292-311.

213. Muir, J.; Gifford, T., *John Muir: His life and letters and other writings*. The Mountaineers Books: 1996.

ABOUT THE AUTHOR

≈ ≈

DR. M JACKSON is a geographer and glaciologist, National Geographic Society Explorer, TED Fellow, veteran three-time U.S. Fulbright Scholar, including two Fulbright-National Science Foundation Arctic research grants, and author of *While Glaciers Slept: Being Human in a Time of Climate Change* (2015). M writes about glaciers and people worldwide and lives outside of Eugene, Oregon. Visit the author's website at: https://www.drmjackson.com.